New Silk Road

The Architecture of the Belt and Road Initiative

Michele Bonino and Francesco Carota with Sohrab Ahmed Marri

New Silk Road

The Architecture of the Belt and Road Initiative

Contributions by Francesca Governa, Stefano Mondozzi, Giulia Montanaro, Lidia Preti, Charlie Xue

Drawings by Sofia Leoni

Photography by CreatAr Images, Ivo Tavares Studio, Raul Ariano, Al Yousuf

Birkhäuser
Basel

Introduction

I. The Architecture of the Belt and Road Initiative: A New Architectural Order?

II. Architectural Guide to the Belt and Road Initiative

Appendix

The BRI Typologies

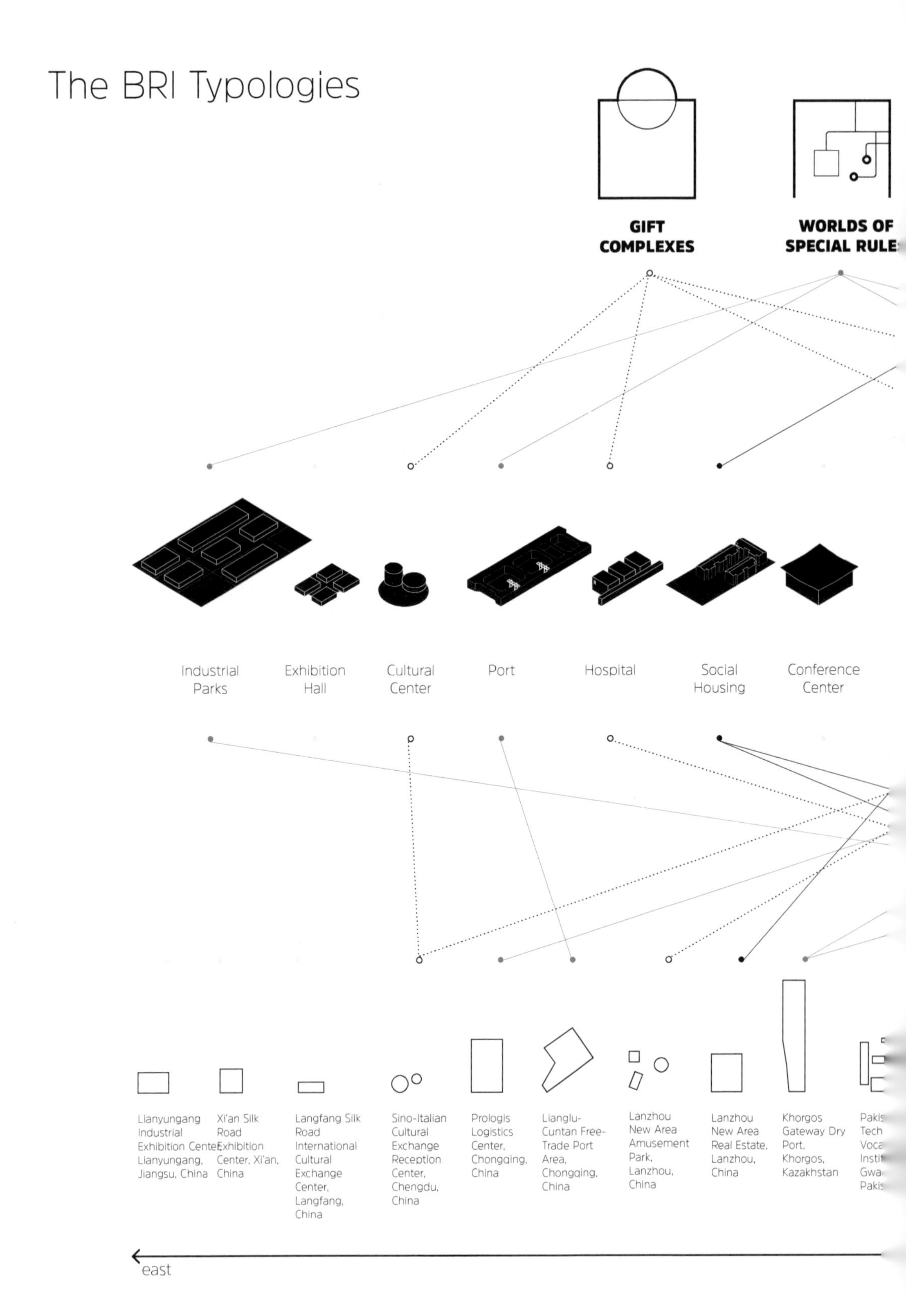

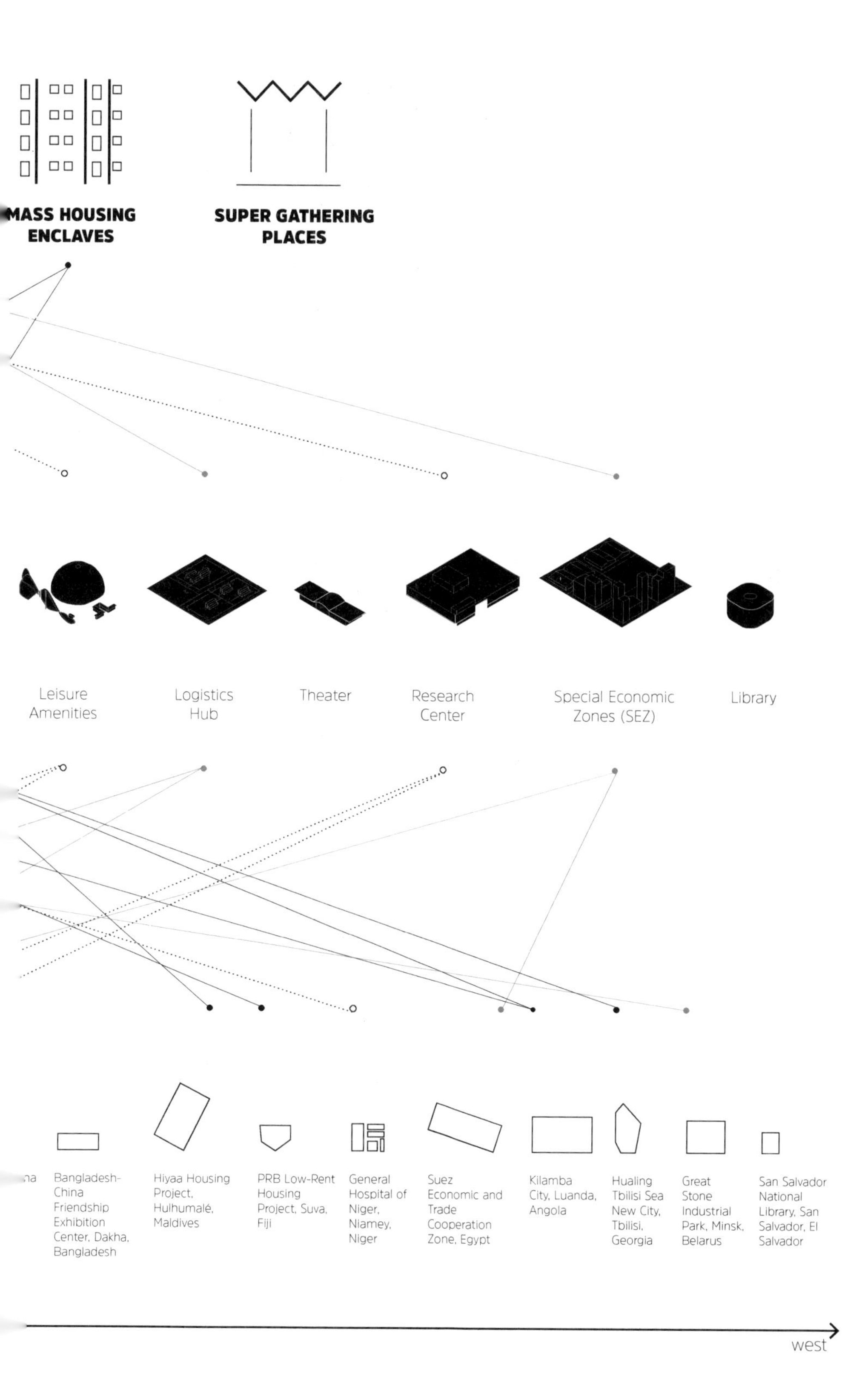
MASS HOUSING ENCLAVES
SUPER GATHERING PLACES
Leisure Amenities
Logistics Hub
Theater
Research Center
Special Economic Zones (SEZ)
Library
Bangladesh-China Friendship Exhibition Center, Dakha, Bangladesh
Hiyaa Housing Project, Hulhumalé, Maldives
PRB Low-Rent Housing Project, Suva, Fiji
General Hospital of Niger, Niamey, Niger
Suez Economic and Trade Cooperation Zone, Egypt
Kilamba City, Luanda, Angola
Hualing Tbilisi Sea New City, Tbilisi, Georgia
Great Stone Industrial Park, Minsk, Belarus
San Salvador National Library, San Salvador, El Salvador
west

The BRI Archipelago

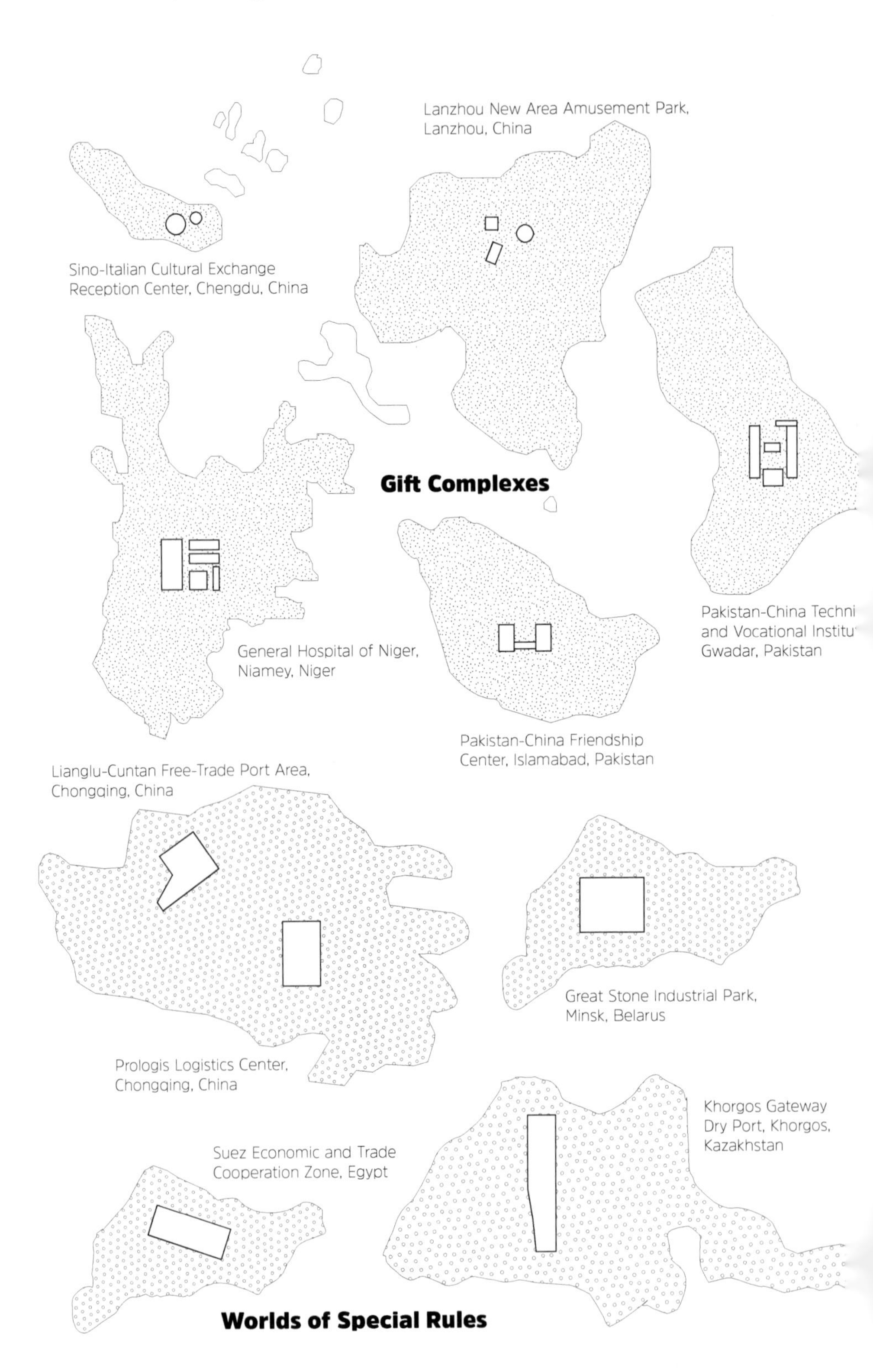

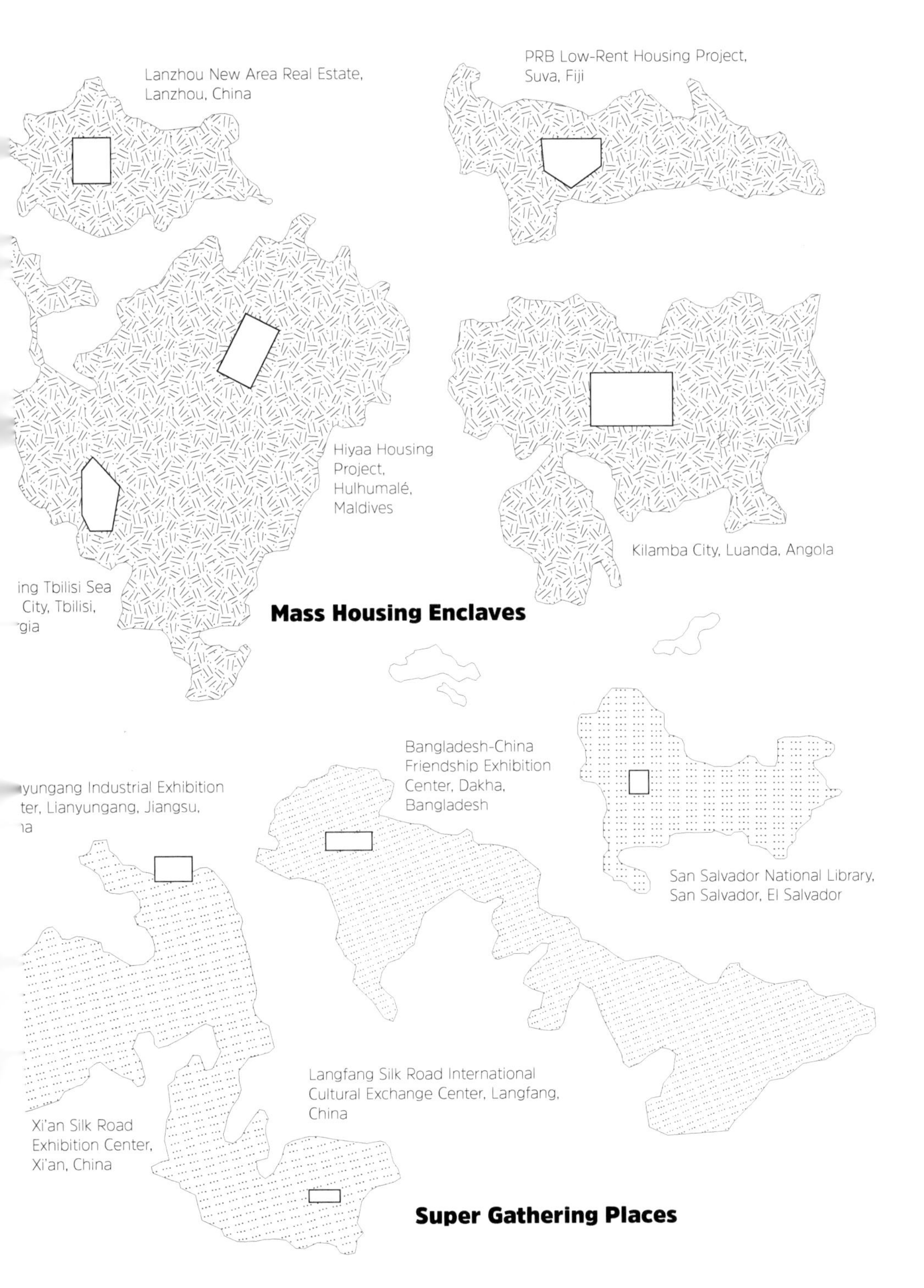
Lanzhou New Area Real Estate, Lanzhou, China
PRB Low-Rent Housing Project, Suva, Fiji
Hiyaa Housing Project, Hulhumalé, Maldives
Kilamba City, Luanda, Angola
ing Tbilisi Sea City, Tbilisi, gia
Mass Housing Enclaves
Bangladesh-China Friendship Exhibition Center, Dakha, Bangladesh
yungang Industrial Exhibition ter, Lianyungang, Jiangsu, na
San Salvador National Library, San Salvador, El Salvador
Langfang Silk Road International Cultural Exchange Center, Langfang, China
Xi'an Silk Road Exhibition Center, Xi'an, China
Super Gathering Places

Introduction

Foreword

Francesca Governa

In late 2013 Chinese President Xi Jinping launched two strategies respectively entitled the Silk Road Economic Belt and the 21st-Century Maritime Silk Road. Together they form the Belt and Road Initiative (BRI), a massive undertaking considered to be the "project of the century" and the world's biggest infrastructure program ever attempted in the last fifty years.[1] By establishing multiple development corridors crossing national borders, the BRI promotes international cooperation via coordinated development policies, new infrastructure systems, increasingly wider investment networks, trade partnerships, and people-to-people bonds.[2] Apart from the official goals, the BRI has other objectives; they include sustaining China's economic growth, bridging the socio-economic gap between China's coastal and the western regions, achieving secure food and energy supplies, consolidating China's leading role among emerging economies, enhancing South–South cooperation, and exporting the "Chinese Dream" as a form of soft power.[3]

However, the BRI does not focus on a single initiative, but involves a plurality of routes, processes, and projects, each varying in complexity and goals. This variety and diversification is not accidental, unintentional, or derived exclusively from the encounter/collision with the ground. Actually, as suggested by Narins and Agnew, the absence of an official map of the BRI is nothing but a "useful fuzziness," so that the uncertainty and malleability of the BRI are its strong point.[4] Indeed, as argued by Murton, the BRI is visualized by "a variety of maps depicting a usefully approximate but inexact network of roads, rails, sea lanes and other infrastructures."[5] Within this array of maps, the issue at stake is their role and "silence," as well as a variety of political and economic aims of the BRI and related projects. The BRI yields, hides, re-emerges and reveals processes which would otherwise be more implicit. In the words of Apostolopoulou: "BRI projects involve almost everything: from railways, airports, ports, pipelines, industrial parks, special economic zones (SEZs), real estate and commercial projects, to free-trade agreements and treaties to boost foreign investment and market liberalization."[6]

The Belt and Road Initiative is now in its second decade, prompting a growing debate that seeks to critically understand today's "global infrastructural turn" and the contemporary "infrastructure scramble"[7] that is rapidly changing vast regions of the planet, from North to South, in Africa, Asia, South America, and Europe. Despite abundant studies and research on the BRI, some issues still require further consideration. Most studies on the BRI adopt geopolitical and geoeconomic perspectives.[8] In geopolitical terms, the BRI is considered as a strategy to globally extend China's political power and influence;[9] from a geoeconomic point of view it is regarded as a means to not only allow Chinese companies to access new markets, but also as a way to address current Chinese economic challenges.[10] However, according to Williams et al., the urban dimensions of the BRI tend to be overlooked.[11] While the rescaling, grounding, and urbanizing of the BRI are now well-established claims,[12] various schools of thought are trying to make sense of them and this requires dialogue between scholars and a comparison between different perspectives.[13] Within the framework of this debate, Dodson believes that "the first aim is to query whether infrastructure is substantially different from other forms of urbanization."[14] This involves: questioning both the BRI and the meaning of urban;[15] exploring the "machinic" aspect of infrastructure;[16] and addressing the always incomplete and uncertain unfolding of urban becoming.[17] In addition, ground-level perspectives have been solicited to "recognize how global initiatives such as the BRI are messy, contingent, and uneven in their outcomes."[18] Other authors highlight the need for inquiries into "Silk Road urbanism,"[19] while Apostolopoulou calls for an investigation regarding the spatial impact of BRI projects and a greater understanding of the conflicting relations between BRI developments and urban transformations.[20]

This book contributes to this growing debate. It also reflects one of the results of a broader research project funded by the Italian Ministry of University and Research (MUR), entitled "Rescaling the Belt and Road Initiative: urbanization processes, innovation patterns and global investments in urban China." Scholars from the Politecnico di Torino (architects, planners, economic and political geographers) and the Università di Macerata (urban economists and experts in Chinese philosophy and culture) were involved in research focusing on the BRI and urbanization and considering their multifaceted material, economic, political and cultural dynamics. The main objective was to address a seemingly obvious issue – the urban role of the BRI – and to try and understand if, and to what extent, a focus on global infrastructure can shed light on the current (global) urban condition. The project studied infrastructure-led urbanization in China, a topic we have explored in previous research projects.[21]

Our research began in September 2020, but its focus shifted due to the Covid-19 pandemic that made traveling impossible. Then new domestic and international tensions arose, for example: geopolitical issues that some scholars critically refer to as the advent of a second Cold War;[22] the wars in the Ukraine and the Middle East; geoeconomic struggles; and the emergence of a multipolar world.[23] Given this complex scenario, we decided to change our perspective. Instead of studying the BRI and the urban in China, we shifted our focus to the BRI implemented beyond China's borders. Our research focused on so-called "Global (urban) China," not only for practical, but also theoretical and epistemological reasons. This shift actually helped us overcome so-called "Chinese exceptionalism" and question the alleged "Chinese otherness."[24] It also integrated BRI studies within the global infrastructure and critical logistics debate. At the same time, it acknowledged that while it is possible to study the imaginaries and material manifestations of the BRI, the latter also involves urban processes – actors, interests, dynamics and so on – that exceed and precede the BRI itself.[25]

Our shifted perspective focused on different localities affected by the BRI;[26] our aim was to closely examine their spatial transformations, the material and socio-technical reconfigurations of infrastructural landscapes, as well as changes in urban geographies along infrastructural networks.[27] In this book we explore something we could call the "architectural turn" of global infrastructure. We ask questions. Does the BRI implement and intensify spatial and architectural transformations? Which architecture materializes the influence of the BRI and China along (variable) infrastructure routes? Is it possible to identify recurrences, describe the forms and functions of these projects, and identify the symbolic, political and economic roles they play? Addressing the BRI as a "materially grounded field of practice,"[28] this book focuses on the specific architecture it produces in different contexts. It adds another layer of interpretation to existing studies on the BRI not only by following the architects and their projects, but also by revealing the transnational networks and flows they are part of and to which they contribute. The book is a guide, uncovering spatial and architectural transformations on the ground and offering a multifaceted, diverse, and transcultural reading of contemporary architecture, thus sparking a dialogue between the places affected by BRI architecture and linking their material and symbolic outcomes to new interpretations and possibilities.

Turin, January 2025

---

1 Emily T. Yeh and Elizabeth Wharton, "Going West and Going Out: Discourses, Migrants, and Models in Chinese Development," *Eurasian Geography and Economics* 57, no. 3 (2016): 286–315; Hong Yu, "Motivation Behind China's 'One Belt, One Road' Initiatives and Establishment of the Asian Infrastructure Investment Bank," *Journal of Contemporary China* 26, no. 105 (2017): 353–368. **2** Ding Fei, "Worlding Developmentalism: China's Economic Zones Within and Beyond Its Border," *Journal of International Development* 29, no. 6 (2017): 825–850; Tim Summers, "Negotiating the Boundaries of China's Belt and Road Initiative," *Environment and Planning C: Politics and Space* 38, no. 5 (2020): 809–813. **3** Kathryn Furlong, "Geographies of Infrastructure III: Infrastructure with Chinese Characteristics," *Progress in Human Geography* 46, no. 3 (2021): 915–925; Weidong Liu and Michael Dunford, "Inclusive Globalization: Unpacking China's Belt and Road Initiative," *Area Development and Policy* 1, no. 3 (2016): 323–340; Stanley Toops, "Reflections on China's Belt and Road Initiative," *Area Development and Policy* 1, no. 3 (2016): 352–360. **4** Thomas P. Narins and John Agnew, "Missing from the Map: Chinese Exceptionalism, Sovereignty Regimes and the Belt Road Initiative," *Geopolitics* 25, no. 4 (2020): 809–837. **5** Galen Murton, "Power of Blank Spaces: A Critical Cartography of China's Belt and Road Initiative," *Asia Pacific Viewpoint* 62, no. 3 (2021): 274. **6** Elia Apostolopoulou, "Tracing the Links between Infrastructure-Led Development, Urban Transformation, and Inequality in China's Belt and Road Initiative," *Antipode* 53, no. 3 (2021): 832. **7** Miguel Kanai and Seth Schindler, "Peri-Urban Promises of Connectivity: Linking Project-Led Polycentrism to the Infrastructure Scramble," *Environment and Planning A: Economy and Space* 51, no. 2 (2019): 302–322. **8** Peter Cai, *Understanding China's Belt and Road Initiative* Sydney: Lowy Institute for International Policy, 2017; Henry W. Zheng et al., "Interrogating China's Global Urban Presence," *Geopolitics* 28, no. 1 (2021): 310–332. **9** Tim Winter, *Geocultural Power: China's Quest to Revive the Silk Roads for the Twenty-First Century* (Chicago: University of Chicago Press, 2019). **10** Peter Cai, *Understanding China's Belt and Road Initiative* (Sydney: Lowy Institute for International Policy, 2017); Tim Summers, "China's 'New Silk Roads': Sub-National Regions and Networks of Global Political Economy," *Third World Quarterly* 37, no. 9 (2016): 1628–1643; Emily T. Yeh and Elizabeth Wharton, "Going West and Going Out: Discourses, Migrants, and Models in Chinese Development," *Eurasian Geography and Economics* 57, no. 3 (2016): 286–315. **11** Joe Williams, Caitlin Robinson, and Stefan Bouzarovski, "China's Belt and Road Initiative and the Emerging Geographies of Global Urbanisation," *The Geographical Journal* 186, no. 1 (2020): 128–140. **12** Gustavo de L. T. Oliveira et al., "China's Belt and Road Initiative: Views from the Ground," *Political Geography* 82 (2020): 1–4; Tim Oakes, "The Belt and Road as Method: Geopolitics, Technopolitics and Power through an Infrastructure Lens," *Asia Pacific Viewpoint* 62, no. 3 (2021): 281–285; Miguel Kanai and Seth Schindler, "Peri-Urban Promises of Connectivity: Linking Project-Led Polycentrism to the Infrastructure Scramble," *Environment and Planning A: Economy and Space* 51, no. 2 (2019): 302–322; Seth Schindler and Miguel Kanai, "Infrastructure-Led Development and the Peri-Urban Question: Furthering Crossover Comparisons," *Urban Studies* 59, no. 8 (2022): 1597–1617; Astrid Safina, Leonardo Ramondetti, and Francesca Governa, "Rescaling the Belt and Road Initiative in Urban China: The Local Complexities of a Global Project," *Area Development and Policy* (2023). **13** Simone Vegliò et al., "A Dialogue on Global Infrastructure-Led Urbanization: Concepts and Reorientations," *Dialogues in Human Geography* (2024). **14** Jago Dodson, "The Global Infrastructure Turn and Urban Practice," *Urban Policy and Research* 35, no. 1 (2017): 90. **15** Tim Bunnell, "BRI and Beyond: Comparative Possibilities of Extended Chinese Urbanisation," *Asia Pacific Viewpoint* 62, no. 3 (2021): 270–273. **16** Ash Amin and Nigel Thrift, *Seeing Like a City* (Cambridge: Polity Press, 2017). **17** Ananya Roy, "What Is Urban About Critical Urban Theory?" *Urban Geography* 37, no. 6 (2016): 810–823. **18** Gustavo de L. T. Oliveira et al., "China's Belt and Road Initiative: Views from the Ground," *Political Geography* 82, no. 2 (2020). **19** Alan Wiig and Jonathan Silver, "Turbulent Presents, Precarious Futures: Urbanization and the Deployment of Global Infrastructure," *Regional Studies* 53, no. 6 (2019): 912–923. **20** Elia Apostolopoulou, "Tracing the Links Between Infrastructure-Led Development, Urban Transformation, and Inequality in China's Belt and Road Initiative," *Antipode* 53, no. 3 (2021): 831–858. **21** Michele Bonino , Francesca Governa, Maria Paolo Repellino, and Angelo Sampieri, eds. *The City After Chinese New Towns: Spaces and Imaginaries from Contemporary Urban China* (Basel: Birkhäuser, 2019); Francesca Governa and Andrea Sampieri, "Urbanisation Processes and New Towns in Contemporary China: A Critical Understanding from a Decentred View," *Urban Studies* 57, no. 2 (2020): 366–382. **22** Seth Schindler, Joshua DiCarlo, and Dipesh Paudel, "The New Cold War and the Rise of the 21st-Century Infrastructure State," *Transactions of the Institute of British Geographers* 47, no. 2 (2022): 331–346. **23** Sandro Mezzadra and Brett Neilson, *The Rest and the West: Capital and Power in a Multipolar World* (London: Verso, 2024). **24** Ivan Franceschini and Nicholas Loubere, *Global China as Method* (Cambridge: Cambridge University Press, 2022). **25** James D. Sidaway et al., "Introduction: Research Agendas Raised by the Belt and Road Initiative," *Environment and Planning C: Politics and Space* 38, no. 5 (2020): 795–802. **26** Other than the places that are already displayed and analyzed in this book, the broader research project focused on Southern Europe, and specifically on the so-called Adriatic Corridor which includes the ports of Piraeus and Trieste. **27** Francesca Governa and Angelo Sampieri, "Infrastrutture Globali e Divenire Urbano: Pireo, Trieste e il Corridoio Adriatico," *Territorio* 103, no. 4 (2023): 23–30. **28** Tim Oakes, "The Belt and Road as Method: Geopolitics, Technopolitics and Power Through an Infrastructure Lens," *Asia Pacific Viewpoint* 62, no. 3 (2021): 281.

Building the New Silk Roads

Francesco Carota and Michele Bonino

There are events, technologies, policies or, more generally, phenomena that can drastically change the world as we know it. Some of them, often less evident than others, are at work in the backgrounds of our lives, bringing radical cultural and physical change to our territories. So, while we are still used to thinking and interpreting the world using our usual frameworks, the environment around us has become so different it appears unintelligible if viewed with a long-established lens.

The Belt and Road Initiative (BRI) is undeniably one of these transformative phenomena. We still don't know the full effects of this plan, but we do know, without a shadow of a doubt, that it represents an important attempt to redefine the vision of our contemporary world. According to the Nobel Prize laureate and economist Joseph Stiglitz, Chinese urbanization is one of the most relevant phenomena we can use to understand human development in the last century; if this is indeed the case, Chinese expansion beyond its domestic boundaries is probably one of the most important and relevant issues that need to be explored in the years to come.

Since its inception by the Chinese government in 2013, the BRI has become a global phenomenon that has so far involved roughly 150 countries worldwide. The program has steadily garnered increasing attention in the media and scholarly discourse, especially in the fields of economics and international relations. In those fields, the BRI is framed primarily as a policy to encourage financial possibilities and political exchanges by strengthening transnational cooperation. Up to now, academic literature dedicated to the BRI is chiefly characterized not only by exhaustive overviews outlining the initiative's effects on global economics, but also by in-depth analyses that seek to reveal the underlying strategic intents, anticipated outcomes, and tactical apparatus employed by China in the propagation of the BRI. And yet, the important footsteps the BRI is materially leaving on the ground in terms of urbanization and architecture appear largely unexplored. Even the infrastructural elements of the BRI that have characterized most foreign Chinese economic investments are not mentioned as physical devices, but as instruments to reinforce connectivity for the sake of collaboration. This approach is as challenging as it is inexplicable: indeed, not only are we not yet fully aware of the enormous system of material and immaterial connections the BRI is building worldwide, but, most importantly, we largely ignore the built landscape they are implementing.

Transnational Infrastructural Spaces

Using the forces of infrastructure to create urbanization was initially an efficient development strategy adopted in China:[1] now, it is a practice exploited worldwide. As discussed by Charlie Xue in his contribution to this book, China's experience in providing construction aid to developing countries is long-standing. It has implemented many important architectural projects as part of a development aid scheme. More recently, however, the scheme has been rebranded into the comprehensive framework of the BRI that now offers cultural exchanges, development projects, and mutually beneficial co-operation, powered by the engine of infrastructure. In other words, when considering the role of China in driving urbanization beyond its borders, this strategy takes on yet another role: to ensure circulation. In this sense, urbanization and infrastructure overlap in a mixture of material and immaterial means which, according to Clare Lyster, are the main forces shaping the contemporary city – a city that is no longer local nor global, but the node of a broader transnational infrastructural landscape.[2] Within the framework of the BRI, infrastructure and urbanization are thus intertwined and reciprocal, not only in the way they depend on and influence each other, but also in the way they bring to life a whole set of new architectural artefacts.[3]

In 2024, as we write this, ten years have passed since the launch of the BRI; in 2026, only two years from now, according to a widely accepted prognosis, China's economic power would reach that of the United States, chiefly thanks to the former's investments in foreign countries. At this moment, we invite architects and urbanists around the world to look more closely at the physical spaces of the New Silk Roads.[4] Examining what the BRI is bringing to our age in spatial and architectural terms is a way of being conscious and inspired about how new urban and infrastructural spaces are shaping our lives. Grounding the BRI, or at least knowing and trying to understand its built spaces, would seem a powerful tool we have as architects to change our point of view on our contemporary practices.

Grounding as a Way of Action

This book analyzes the BRI principally and primarily as an architectural phenomenon tackled on and from the ground. In light of a new materialist approach to critical problems, we aim to establish an understanding of the BRI in which human agency is no longer primary, but instead the centrality of architecture is affirmed as a complex system made of material and/or immaterial elements and interconnected relations, thus defining buildings and urban structures. However, adopting this kind of perspective has important consequences. It is another way of saying that BRI material objects are no longer simply part of universal infrastructure, as they have been mainly conceived in the past, but are objects that alight on different grounds. By this, we do not only mean they belong to different parts of Mother Earth, but that they should and could have different foundations; not only do they have specific cultural, economic, and political foundations, but they also have different points of departure generating systematic architectural knowledge.

This book argues that BRI territories are places and, for this reason, we intend to map the variety and plurality of their architectural outcomes. Drawing on recent scholarships in environmental studies and urban geography,[5] our research is based on the act of "grounding" in order to reveal the important role played by different places and cultures in the comprehension of architecture and urban environments. We have adopted an approach which, from the ground up, examines how architecture deals with –

and is shaped by – specific processes of circulation, colonization, and dispossession. This does not mean, however, that we wish to seek refuge in some form of local and situated knowledge. We are all too aware that current urban architecture, and BRI architecture in particular, is never only local. Instead, it is strongly characterized by the exchange of materials, rules, and practices of human and non-human agents – not only architects, politicians, constructors and developers, but also algorithms, devices, and robots. BRI architecture is part of a transnational network and is considered here in this manner. The issue is not, as for Lefaivre and Tzonis,[6] to promote an alternative to universal, centralized design models driven by global or multinational corporations and institutions, but rather to suggest new ways of reading contemporary architecture as multiple, diverse, and transcultural. In this regard we draw on postcolonial and decolonial scholarships that resist consolidated frameworks, models, and categories and shift the centre of architectural theory and practice to broader territories, using them as a starting point for pursuing a more inclusive understanding of architecture and urbanism. Global architectural discourse tends to be extremely homogeneous; it relies on "architectural knowledge" largely based on information from the Global North. The BRI provides an interesting opportunity, i.e., to exploit the growing number of architectural projects in the Global South. While we have extensive literature about North-South relations in architecture and urban planning – ranging from colonial times to the post-war period and contemporary globalization – the BRI presents new global relationships occurring mainly in East-South relations.

However, rediscovering the foundation behind diverse architectures is a meaningful activity if, as architects, we are able to translate this knowledge and way of thinking into our practice. To achieve this goal, we need collective action that can help multiply our voices by creating tension between different representations of contemporary architecture, thus revealing its complexity around the world. That is the reason why this book calls on architects and scholars from the global community to not only participate and contribute to the exploration of the transnational infrastructure spaces of the BRI, but also provides new theoretical lenses with which to interpret the paradigmatic changes involved. The book's goal is to establish a basis for further contributions that wish to walk along the challenging path of laying the groundwork for a new architecture or, in other words, to "ground" the BRI. It is a guide to encourage new explorations – an objective outlined in the final pages of this book.

A Guide as an Overture of Possibilities

If, according to Rem Koolhaas,[7] Manhattan needed a "Manifesto," the Belt and Road Initiative probably needs a "Guide." So, as scattered and unknown as it is, the BRI requires new frameworks and voices if we are to understand its material and social implications on the ground. There are many reasons why, as architects and non-architects alike, we are interested in buildings. However, when we look at buildings and urban environments, they can often appear almost meaningless. They transmit no feelings, no emotions, and no logic. This is especially true when they are unfamiliar; just like an unfamiliar piece of music might leave us cold the first time we hear it. The information in an architectural guide would presumably enhance our architectural experience. So hopefully this book will help scholars, practitioners, and students to view and appreciate the architecture of the BRI with new eyes, as well as pave the way for more detailed investigations vis-à-vis this phenomenon.

Compiling a guidebook is not just an objective task. It is raw and critical in itself. Firstly, a guide is selective, and selection is a fundamental step behind its conception. So, how is it possible to choose from among a wide range of projects built worldwide? To some extent the BRI did it for us. Irrespective of whether the BRI is the primary driver for developments carried out in its name, according to Tim Bunnell it is a label that places brand-new infrastructural developments and previously unassociated project sites in a comparative relationship.[8] However, working as an almost random label, at least in architectural terms, the BRI is an overture to possible intricate interchanges and associations. Its name can be applied to a dry port in Khorgos, a new hospital in Niger, or a social housing development in Africa; its label covers everything built within its financial boundaries, apparently without restrictions in terms of typology, architecture, and built form.

We can say that Chinese pragmatism has led to extreme architectural variety based on an intrinsic and specific political logic; variety is, therefore, one of the essential characteristics of BRI architecture: a basically disengaged approach appears to welcome forms and imaginaries of local historicism without major problems. On the whole, it seems to configure a new form of cultural influence, no longer based on the export of consolidated models and styles - a common approach by Western powers in earlier colonialist eras - but on a more inclusive and flexible relationship with formal imaginaries and local repertoires.

The projects documented in this book reveal several different strategies, narratives, and practices specific to the BRI, but also paradigmatic of the patterns of logic, rules, and standards that are forming a new doctrine of architectural development, one which can transcend the specificity of its financial origin. The architecture of the BRI is, indeed, an example par excellence of thousands of industrial parks, logistics centers, housing communities, conference centers and other facilities built worldwide, driven by this new Chinese-centered urbanism. Understanding the architecture of the BRI, or at least exploring some of its features, involves gathering not only information specific to the initiative, but also general insights into the trajectory of contemporary architecture, starting with specific and local data.

Nevertheless, for these very reasons, the examples presented here are but a small selection. This book documents just a few of the many existing BRI-related projects. Several of them have been built, some are still in the planning stage, some are small and restricted interventions, while others trigger major changes to the surrounding territories; all of them, however, have been co-promoted, co-designed, and/or co-built by one Chinese entity. Knowledge of this showcased architecture is not intended to be exhaustive as regards the selection, or comprehensive in its analysis. Nevertheless, its aim is to provide a starting point from which to try and understand the BRI in architectural terms. What is important is that, from port cities to social housing or gated communities, this selection of multiple building types also reframes our understanding of what we consider valuable architecture. The "Pevsner Architectural Guides. The Buildings of England"[9] included in its selection many ordinary offices, farms, and even silos, thus encouraging a reconsideration of what forms of architecture were relevant at that time. Likewise, our selection challenges the common understanding of the limits of what constitutes "architecture." In this regard, the objective we wish to achieve with our guide is to provoke an ideological shift among practitioners and academics, urging them to rethink the value of some of the pragmatic and utilitarian works that often fly under the architectural radar.

Finally, like all architecture guides, this book entails descriptive investigations, supported by custom-made maps, drawings, and diagrams, thus allowing the reader and observer to develop their own interpretations and conclusions. It also provides interpretive taxonomies in order to understand the architecture of the Belt and Road Initiative. After identifying and listing a large number of buildings and urban artifacts, making categories became almost natural. Categorization is a fundamental process of human intelligence as well as a key factor in investigations and research in the field of architectural science. Categorization helps us reduce complexity and group objects thematically. The guide acts as an interface; it reveals and, at the same time, conceals and excludes. After having analyzed many case studies in detail, we propose a preliminary reading of the BRI architecture, starting with four groups.

Gift Complexes. Hybridizing Transnational Architecture. The BRI is primarily narrated as an infrastructure for business growth, but other urban materials arise when business activities touch the ground and come into contact with communities. The leisure, commercial, and medical spaces donated by China to other countries are indirect interventions in the financial strategies of the BRI; due to their eccentric and media-oriented architectural language, their aim is to distract both local inhabitants and the global audience from politically and socially controversial issues.

Worlds of Special Rules. Architectures between Humans and Data-Driven Machines. The BRI is a good opportunity to treat logistics and automation as an architectural issue, or at least a spatial one. Free-trade zones, logistics hubs, and industrial parks featuring the most innovative and surprising solutions transform the ground of the New Silk Roads into a machinic landscape where flows of people and goods help to shape the built environment, considering the political problems of logistics as fundamentally architectural. Often these zones are either extraterritorial or their access is tightly controlled and restricted – hence they are realms governed by special rules.

Mass Housing Enclaves. Between Standard Forms and Local Conditions. Since the geographies of the BRI feature variegated and sometimes extreme territories, ranging from deserts to artificial islands, identifying housing in those geographies is not always an easy task. However, most of the living spaces provided involve mass-produced compounds that rarely differ from one place to another in terms of form, programs, and construction techniques: they represent the housing solution for a large slice of the world's urban population, as well as a housing dream for many others. While there is a tendency to produce a standardized roll-out, some adaptation to local scenarios can be observed.

Super Gathering Places. Meeting in Between Architecture. Spaces designed to host meetings, exhibitions, and trade fairs are typically over-emphasized in the BRI discourse, portraying the initiative primarily as a strategy of relations, discourses, and diplomacy. Thanks to structural innovations and spatial grandeur, these buildings are key drivers behind the business opportunities and exchanges that arise on the ground.

On the one hand, these four groups reveal several unique features of BRI architecture, while on the other they open up new avenues to discuss this kind of architecture based on well-established interpretive lenses: respectively "language," "distribution, "standardization" and "tectonics." Instead of being monolithic frameworks, these categories become points of departure to explore the tensions between different and opposite interpretations.

In Need of a Non-Dichotomous Architecture

In this book we are not proposing anything new: modern and contemporary architects have always observed the existing architectural landscape and used it to understand the transformative forces of their age. Le Corbusier was fascinated by primitive industrial structures; Venturi, Izenour, and Scott Brown[10] chiefly delved into the ordinary American landscape to fight the creeds of modern architecture; Rem Koolhaas[11] theorized the early development of Manhattan as a way to reframe the foundations of the contemporary metropolis. These are just a few examples; Chinese-based urbanization around the world – with the BRI as its most obvious expression – might simply be another one.

Numerous dichotomies emerge while examining the epistemological ground of the architectural discipline. On the one hand, dichotomies are indeed a way of structuring our thoughts; on the other, they can be problematic. This is particularly evident when we establish relations between big theoretical frameworks and specific outputs on the ground, as illustrated by the afore mentioned studies. Salient divisions and dichotomies can even become additional conceptual constraints due to the growing complexity and geographical expansion of the architectural field, coupled with the proliferation of new forms and modes of practicing. When we meticulously examined the architectural endeavors within the BRI we often started by using specific lenses; this led to numerous dichotomies, such as generic/specific, global/regional, form/ornamentation, and control/freedom. However, we often ended up somewhere else, faced with an architectural landscape that was primarily generated through negotiation and diplomacy. Indeed, despite their very diverse contexts and approaches, the case studies presented in this guide share one discernible feature: a nuanced sense of neutrality, a lack of hegemonic imposition, and thus an intrinsic penchant for shaping diverse and not dichotomous architectural orientations and aspirations.

While discussing these interplays and how they contribute to the blurring of established dichotomous assumptions in the field of architecture, this book's objective is neither to propose a new way of looking at architecture, nor provide specific answers; on the contrary, its goal is to open up new questions. The book does not aim to be conclusive, but rather to initiate a journey. This publication represents our efforts to trigger an incremental process to gather together not only BRI-related projects, but all the new architectural endeavors that come from, and are imposed on, diverse territories; projects and endeavors that can challenge dichotomous constraints and establish multiple trajectories in order to understand the contemporary architecture of transnational infrastructural spaces.

1 See, from the authors' research group (China Room, Politecnico di Torino), Michele Bonino, Francesca Governa, Maria Paola Repellino, and Angelo Sampieri, eds. *The City after Chinese New Towns.* (Basel: Birkhäuser, 2019); Michele Bonino, Francesco Carota, Francesca Governa, and Samuele Pellecchia, eds. *China Goes Urban. The City to Come.* (Milan: Skira, 2020); Leonardo Ramondetti, *The Enriched Field, Urbanising the Central Plains of China.* (Basel: Birkhäuser, 2022). **2** Clare Lyster, *Learning from Logistics.* (Basel: Birkhäuser, 2016). **3** For an early example of Chinese-built infrastructure in Africa, see for instance Sascha Delz, "Who Built This? China, China, China! Expanding the Chinese Economy Through Mutual Benefit and Infrastructure Construction," in Marc Angélil and Dirk Hebel, eds. *Cities of Change: Addis Ababa. Transformation Strategies for Urban Territories in the 21st Century.* Second and revised edition. (Basel: Birkhäuser, 2016): 198–206. **4** Even though commonly described as the "New Silk Road," the program features not just one but multiple development corridors crossing national borders. **5** See for instance Ananya Roy and Aihwa Ong, eds. *Worlding Cities: Asian Experiments and the Art of Being Global.* Studies in Urban and Social Change. (Malden, MA: Wiley-Blackwell, 2011); Henrik Ernstson and Sverker Sorlin, eds. *Grounding Urban Natures: Histories and Futures of Urban Ecologies.* (Cambridge, MA: The MIT Press, 2019). **6** Liane Lefaivre and Alexander Tzonis, *Critical Regionalism: Architecture and Identity in a Globalized World.* (Munich: Prestel Pub, 2003). **7** Rem Koolhaas, *Delirious New York: A Retroactive Manifesto for Manhattan.* (New York: The Monacelli Press, 1997). **8** Tim Bunnell, "BRI and Beyond: Comparative Possibilities of Extended Chinese Urbanisation." *Asia Pacific Viewpoint* 62, no. 3 (2021): 270–7w3. **9** Nikolaus Pevsner, *Pevsner Architectural Guides. The Buildings of England.* (London: Penguin Books, 1951). The series started in 1951 and eventually ran to 46 volumes. **10** Robert Venturi, Denise Scott Brown, and Steven Izenour, *Learning from Las Vegas.* (Cambridge, MA: The MIT Press, 1972). **11** Rem Koolhaas, *Delirious New York: A Retroactive Manifesto for Manhattan.* (New York: The Monacelli Press, 1997).

The BRI before the BRI: China's Architectural Aid and International Engagement

Charlie Xue

Even though the Chinese government proposed the Silk Road Economic Belt and 21st-Century Maritime Silk Road Initiative in 2013 as an innovative conceptual and practical instrument to promote exchange and economic development, the initiative did not appear spontaneously. Its origins can be traced back to China's extensive international engagements over the preceding six decades. Throughout the Cold War, from 1950 to 1990, and the country's globalization era starting in the early nineties, China never stopped developing relationships with other countries by contributing foreign aid. The underlying belief was that fostering stable and peaceful international relations was pivotal to ensuring the prosperity and well-being of the nation and its citizens.

While cultivating international ties, China extended its goodwill to numerous developing countries in Asia and Africa by offering foreign aid in various forms. These programs were implemented through construction projects, materials, technical cooperation, collaborative human resources development, foreign medical teams, emergency humanitarian aid, volunteers, and debt relief.[1] Among these, construction projects stood out as the most visible and long-lasting in recipient countries. During the sixty-year period from 1958 to 2018, the construction by Chinese professionals of over 1,500 structures (including railways, stations, factories, congress and assembly halls, theaters, stadiums, schools, hospitals, and government buildings) was subsidized in over 160 countries by the Chinese government that provided other forms of economic aid for agriculture, irrigation, education, transportation, manufacturing, and health, i.e., investments totaling more than US$10 trillion.[2]

Why, and under what conditions, were Chinese-funded buildings proposed, designed, and constructed over a period of more than fifty years? Did these buildings promote social progress and improve people's lives in developing countries? How did these buildings enrich the modern architectural movement outside Western countries? What are the main differences between donated and commercial buildings? This chapter briefly addresses these questions and explores the background of the BRI by discussing the reasons behind architectural export, a tool directly or indirectly associated with complex social, political, economic, cultural, and ideological factors. It also describes China's foreign aid construction projects during three periods: the early socialist years led by Mao Zedong from the fifties to the seventies, the "open door" era in the eighties and nineties, and the "going out" era in the 21st century. Providing a summary of the construction projects undertaken by Chinese professionals overseas contributes to the study of cross-border design and modernist architecture outside of Western countries.

Early Socialist Years: The 1950s to 1970s

Soon after the establishment of the People's Republic of China, the socialist camp started to provide it with foreign aid for mass construction projects and technical support, thus helping to shape the industrial layout of the new China.[3] These initiatives played a pivotal role in shaping the industrial landscape of China, with assistance from the Soviet Union and Eastern European nations laying the groundwork for China's heavy industry, education, and culture. It brought about an era of learning in all fields from the Soviet Union, China's "big brother." Visual and material reminders of this aid include the Sino-Soviet Union friendship buildings, built in Beijing and Shanghai in 1956, and the vehicle factory in Changchun, constructed just one year earlier.

In early 1955, despite narrowly escaping assassination while en route to the Bandung Conference of the Non-Alliance Movement, Chinese Premier Zhou Enlai (1898–1976) participated in the event; by promoting his Five Principles of Peaceful Coexistence he facilitated the drafting of a concluding declaration that allayed the doubts and fears of several anticommunist delegates about China's intentions.[4] Indeed, shortly after the conference, China established diplomatic relations with more than twenty Asian and African countries and began to not only increase its trade activity, but also to expand its foreign aid.

In the early sixties, when China was surrounded by enemies during the Cold War, it attempted to break through the diplomatic blockage by funding construction projects in non-aligned countries in Asia and Africa. China's donations reflected the philosophy of the country's top leaders, a philosophy clearly expressed by Chairman Mao Zedong (1893–1976) in the following statement: "Since China has a territory of 9.6 million square kilometers and 600 million people, our country should contribute more to mankind"[5] and "those who gained the revolutionary victory should help other people who are still struggling for liberation. This is our international responsibility."[6] From that moment on, Mao's words were manifest in China's foreign aid policy throughout his regime.

China's foreign aid was given chiefly to newly independent countries and countries holding similar socialist ideals.[7] The projects were all designed by Chinese architects and primarily constructed by Chinese companies. The blueprints were usually produced in Chinese and the construction materials, facilities, furniture, and even cutlery, were shipped from China.

The fifties and sixties were a time of economic stringency in China. Famine ravaged the country, resulting in the death of millions.[8] At the time most civil buildings were built on a tight budget. However, foreign aid construction projects were usually higher in quality. As reported by many sources,[9] these projects were an excellent field of experimentation for contemporary Chinese architects. They were provided with valuable opportunities to practice modernist principles and express regional tastes, which were banned in socialist China at that time. During the Cultural Revolution (1966–1976), when domestic construction had, by and large, ground to a standstill, China's

Bandaranaike Memorial International Conference Hall, Colombo, Sri Lanka, 1973.

foreign aid programs incorporated annual large-scale construction projects. These projects included stadiums, international conference halls, and infrastructures, such as the 1,860-kilometer long Tanzania–Zambia Railway (1975), and more than 100 stations, depots, and workshops. By the end of the Cultural Revolution, foreign aid accounted for 7% of China's national expenditure; this was not financially sustainable.[10]

During the Cultural Revolution, China built factories, conference centers, and a few stadiums, mainly overseas. These two building types were frequently constructed in China in the sixties and seventies. Many of these factories in Africa and Asia are now dilapidated after decades of use and wear. Some conference centers, including the Bandaranaike Memorial International Conference Hall in Sri Lanka (1973), the People's Palace in Guinea (1967), and the Friendship Hall in Sudan (1976), are still a source of national pride, especially after China's renovation efforts in the 21st century. Drawing on their experience in building the Great Hall of the People in Beijing and other Chinese provincial cities, Chinese architects designed these conference centers incorporating aspects of tropical modernism. In particular, Chinese architects researched the building technology and design language used in the hot climate of the Global South and proceeded to create buildings in a different manner compared to the method used in China.[11]

In the early fifties China began to nationalize all private design firms. An architectural professional could either work at a state-owned design institute or teach at a university.[12] State-owned design institutes were the 'instruments' of the central, provincial, and municipal governments. Designing a foreign aid construction project was therefore assigned to state-owned design institutes, which often considered this task to be an honor; the projects were often assigned to the institutes' best designers. The designers flew to the designated site, worked on the design for several months, and then illustrated the projects to the leaders of China and the host countries. This extended timeframe facilitated innovative thinking, discussion, comparison, and elaboration, thus ensuring high-quality designs.

Foreign Aid During the Open Door Era: The 1980s to 1990s

China's announcement of the "open door policy" in 1978 was a turning point; it led to a construction boom within the country and a wave of foreign architectural influences. "Open-door" welcomed foreign investment and the importation of technology and management, and inevitably brought foreign cultures and lifestyles to mainland China. By 2013, when the BRI was launched, design competitions and invitations had resulted in the construction of over 700 landmark buildings in mainland Chinese cities, all designed by foreign architects.[13] These construction projects enabled Chinese architects to interiorize the design methods and technologies used by their Western peers and apply them in their architectural designs for foreign aid construction projects.

In the last twenty years of the 20th century, the number of China's foreign aid construction projects increased considerably, and the host areas expanded from Asia and Africa to Oceania and Latin America. In addition to conference centers, the projects included indoor and outdoor stadiums, schools, hospitals, theaters, and youth centers. Due to China's rapid economic growth, there was a drop in the proportion of China's expenditure allocated to foreign aid. During the Mao era, the design and construction of foreign aid projects was assigned to large, state-owned companies. During the "open door" era, designs were selected based on competitions and construction contracts were obtained through bidding. This provided new opportunities for firms other than large, state-owned design institutes.

Outstanding projects emerged within this dynamic landscape, reflecting the era's hallmark commitment to exceptional design and construction quality. Examples included the Cairo International Convention Centre in Egypt, designed by the Shanghai Institute of Civil Architectural Design (completed in 1990), and the Moi International Sports Centre in Nairobi, Kenya, designed by the China Southwest Design Institute of Architecture (completed in 1987). During this period, sports venues and facilities were among the most common building type constructed as part of China's foreign aid programs. By 2019, over 110 stadiums in more than seventy countries had been completed as part of China's foreign aid construction projects.[14] This impetus boosted China's contribution to the construction of soccer stadiums, indoor stadiums, swimming pools, and athlete hostels in preparation for the African Cup of Nations soccer competition and the Asian Games. Some of the stadiums that China helped to build, for example the 40,000-seat stadium constructed in Dakar, Senegal, in 1996, have even greater seating capacities vis-à-vis comparable stadiums in China.

Other examples of changes in the profiles of architectural firms engaged in overseas design activities were implemented by the design institute in Hangzhou, led by Cheng

Chinese design team in front of the Cairo International Convention Centre, 1989.

Taining – tasked with designing the National Theatre of Ghana, completed in 1991 – and by a design institute in Guangxi, charged with the design of a theater in Myanmar in 1986; after hosting numerous performances, the theater was renovated in the first decade of the 21st century. These works reflect the ability of Chinese province-level and city-level (second-tier) design firms (rather than national firms) to produce high-quality designs.

Moreover, in the eighties and nineties the architectural designs of China's foreign aid construction projects were developed considering regional characteristics. This trend was in part influenced by architectural concerns in China, such as postmodern and vernacular architecture, and the principle of focusing on location and context. Most buildings were constructed in tropical areas, where designers had to use special strategies to combat the challenges posed by heat and humidity. Tropical architectural concepts and modernism strongly influenced China's domestic construction projects in the late seventies to eighties. In the eighties and nineties, China transitioned from a planned economy to a market economy. After the turbulent years of the Cultural Revolution, architects were strongly motivated by the existing free and creative atmosphere and enthusiastically 'swam' in the sea of the market economy. This encouraged incorporating diverse, innovative ideas into the architectural designs for various types of foreign aid construction projects.

"Going Out" in the 21st Century

At the dawn of the new millennium, China embarked on a new chapter in its global engagement, signaling an era of globalization. Beijing launched the Forum on China–

National Theatre in Ghana, 1991.

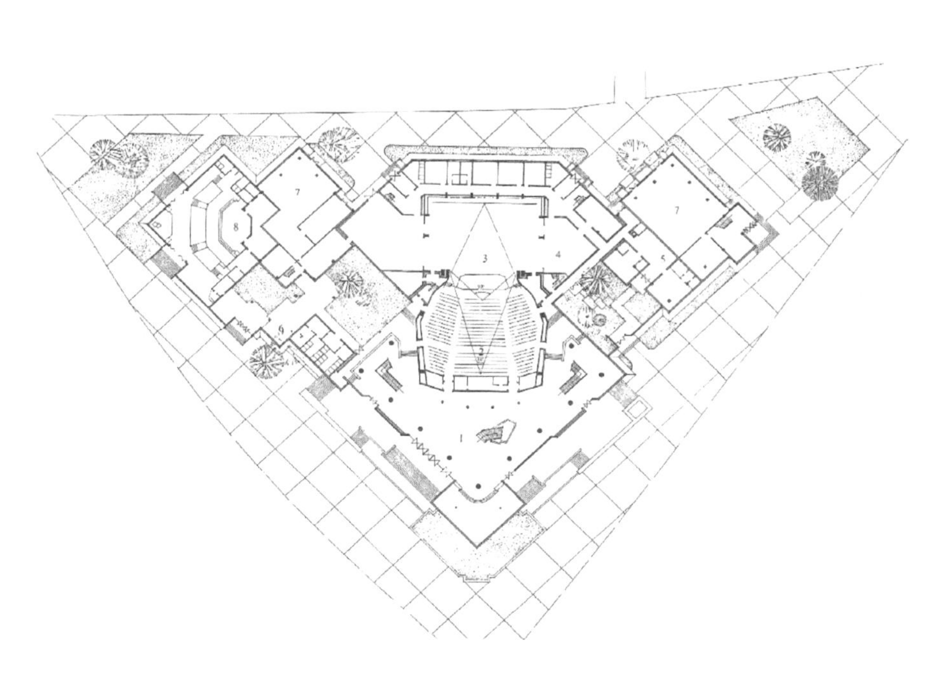

Plan of the National Theatre in Ghana, 1991.

Africa Cooperation in 2000, and China joined the World Trade Organization in 2001. Hosting the Olympic Games in 2008 and the World Expo in 2010 bestowed momentum on China's efforts to integrate into the international community. According to AidData,[15] from 2013 to 2018 China contributed a total of RMB 270.2 billion (US$41 billion) in foreign aid to countries in Asia (30), Africa (53), Oceania (9), Latin America and the Caribbean (22), and Europe (8).

Since the start of the "going out" era, many state-owned and private companies have looked for commercial opportunities overseas. Foreign aid projects have helped Chinese companies to "go out" (i.e., invest and obtain work overseas), engage in economic and cultural exchange, and achieve mutually beneficial foreign partnerships.[16]

Since the beginning of the "going out" era, China's construction projects have included foreign aid donations and commercial projects financed by Chinese banks or foreign investors. Design and construction companies acquire commissions through bidding and, in most cases, engineering, procurement, and construction (EPC) contracts. These contracts cover the design, construction, cost control, and services, often necessitating adherence to the building codes and standards of the host countries, thereby introducing additional challenges into the design process.

Surprisingly enough, in the first two decades of the 21st century China undertook five times as many foreign aid construction projects as it did throughout the 20th century.

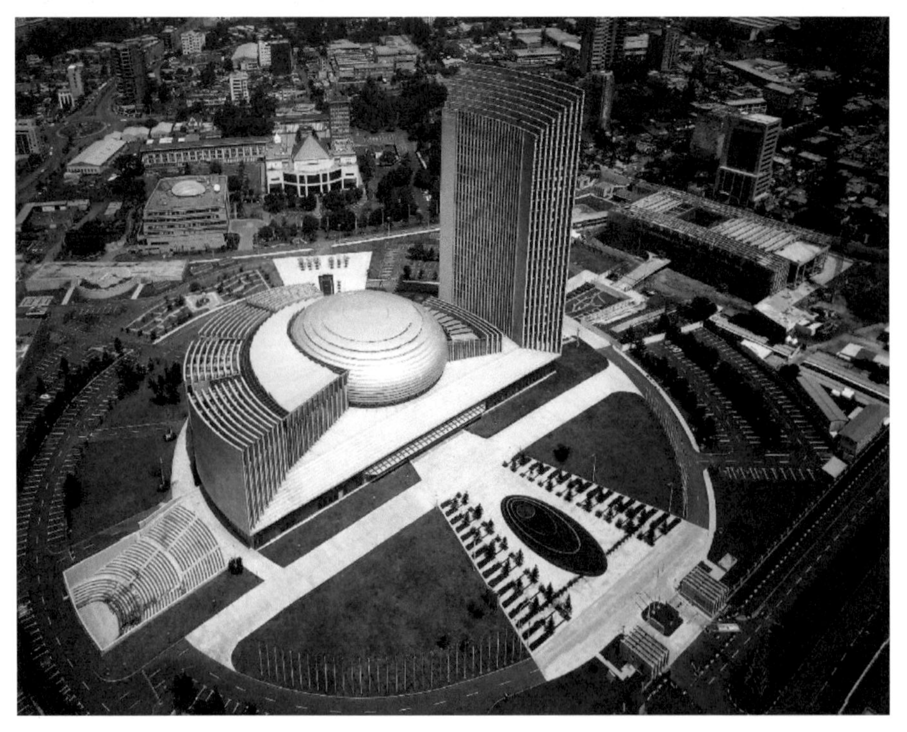

African Union Headquarters, Addis Ababa, 2011.

This surge translated into a profusion of opportunities for design and construction companies. However, registration with the Ministry of Commerce is a prerequisite for design institutes seeking involvement in foreign aid initiatives. Beijing, Shanghai, and Wuhan have historically been home to large, well-established, state-owned design institutes, i.e., they are the primary contenders in China's architectural design industry. However, new competitors have emerged from other channels and in the private sector. For example, China IPPR International Engineering was originally a design institute managed by the Ministry of Mechanical Engineering. In the seventies the company began to undertake foreign aid projects involving factory design and construction. Since then, the company has completed approximately 200 projects, including the construction of hospitals, schools, office buildings, stadiums, conference centers, and industrial buildings in more than fifty countries in Asia, Africa, Latin America, Europe, and Oceania. More than forty hospitals designed by IPPR are open to people in developing countries. The company provides not only the design, but also services related to feasibility studies, project management, and contracting.

In China, a design scheme prepared by a listed design firm must first win a design competition before it can implement its project. In many cases, the leaders and local juries in host countries participate in the selection process, attracted by designs that exude architectural charisma and have a talent for "storytelling." For instance, IPPR's designs for the Senegal national wrestling arena (2018) and the Morodok Techo National Stadium in Cambodia (2021) accommodated sophisticated functional requirements and highlighted the national symbols of the host countries; the large truss in the Senegal wrestling arena symbolizes the golden belt of a wrestler and the 100-meter-high structural portal in the Cambodian national stadium symbolizes a traditional greeting (namaste, a friendly gesture).

China's "going out" policy and the BRI have not only facilitated the construction of high-quality stadiums, theaters, and conference centers in developing countries, but have also provided opportunities for China's private developers and designers. Private developers can use the market to identify suitable projects, such as mass housing estates, shopping malls, hotels, or entertainment venues. When private developers invest in a foreign project, they can hire affiliated professionals overseas to work on it.

Other compelling examples of how China's "going out" policy has influenced design construction in other countries are the shopping mall and housing estates built in Tbilisi, Georgia by Hualing, a real-estate group based in Xinjiang. Encouraged by China's "going out" policy, Hualing drafted blueprints for a new district in Georgia starting in 2007; work began within the framework of the BRI after 2014.

In addition to private developers, private design firms find their clientele directly in overseas markets. For example CCDI, a firm that was established during the preparations for the Olympic Games in the early 21st century, has consistently undertaken projects in the Middle East, Asia, and Africa. There are three ways the company acquires overseas projects: by cooperating with construction companies to bid on EPC

Wrestling Arena, Senegal, 2018.

Sea Plaza, Tbilisi, Georgia, 2017.

contracts; by purchasing competent design firms in foreign countries (e.g., Australia and the United States); and by directly providing services to Chinese developers who invest overseas.[17]

Conclusions

This chapter has provided a brief overview of China's foreign aid construction projects and the architectural operations implemented in three historical periods. During the Mao era, the task of designing buildings abroad was chiefly assigned to large state-owned design institutes; their designers applied their skills to the assigned tasks, experimenting with standards that were higher than the ones applied in comparable construction projects in China. During the Open Door era, Chinese firms obtained design commissions through competition-based design selection and obtained construction contracts through bidding; this helped to improve quality and control the cost of these construction projects, making them more varied in terms of building typology and architectural principles. In the early 21st century, Chinese design firms started to consider the design of both foreign aid and commercial construction projects as a profit-earning activity to be performed under tight time constraints and strong market pressure; this led to an increase in the number of projects implemented by mixing repeatable Chinese protocols and technologies and marketable local features.

In short, China's foreign aid construction projects represent a long-term commitment to developing countries and embody multiple meanings beyond their architectural aesthetics. The buildings physically reflect the complex relationships that exist between China and other countries. Foreign countries regard aid construction projects as symbols of power and legitimacy, as well as opportunities to develop technology and artisanship through construction. On the other hand, China's foreign aid construction projects have been heavily influenced by domestic ideologies, politics, culture, economics, and technology. They partially reflect the changes that have taken place in Chinese society, incorporating aspects of tradition and modernity. Chinese architects integrate modernist principles and distinct Chinese experiences with characteristics of the local environment.

Since the start of the BRI, more Chinese design and construction companies have been turning their attention toward overseas investment and showing off their talent by bidding on international contracts. Overseas design projects have become part of the normal business of those firms. In the past decade more foreign countries have expressed their desire to participate in China's foreign aid construction projects and promote collaboration between local and international designers. Chinese firms, facing more competition at home and abroad, have gradually enhanced the quality as well as the cultural and political significance of foreign aid construction projects, thus leading to new efforts in the innovation and disruption of architectural means.

---

1 State Council of China. *White Book on China's Foreign Aid.* (Beijing: State Council of China, 2011). **2** For a comprehensive review, see Lloyd Amoah, "China, Architecture and Ghana's Spaces: Concrete Signs of a Soft Chinese Imperium?" *Journal of Asian and African Studies* 51, no. 2 (2016): 238-255; Deborah Brautigam, *The Dragon's Gift: The Real Story of China in Africa.* (New York: Oxford University Press, 2011). **3** Hua Gao and Liu He, *Zhonghua Renmin Guoheguo Shi [A History of the People's Republic of China].* (Beijing: Zhongguo Dangan Chuban She, 1995). **4** A. Doak Barnett, *Communist China: The Early Years, 1949-55.* (London: Pall Mall Press, 1964). **5** Zedong Mao, *Selected Works of Mao Zedong.* Vol. 7. (Beijing: People's Press, 1996). **6** Ibid. **7** Wolfgang Bartke, *The Economic Aid of the PR China to Developing and Socialist Countries* (New York: K. G. Saur, 1989); Lloyd Amoah, "China, Architecture and Ghana's Spaces," *Journal of Asian and African Studies* 51, no. 2 (2016): 238–255; Charlie Q. L. Xue et al., "Architecture of 'Stadium Diplomacy' – China-Aid Sport Buildings in Africa," *Habitat International* 90 (2019). **8** Frank Dikötter, *Mao's Great Famine - The History of China's Most Devastating Catastrophe, 1958-62.* (London: Bloomsbury, 2010). **9** See for instance Charlie Q.L. Xue, *Building a Revolution: Chinese Architecture since 1980.* (Hong Kong: Hong Kong University Press, 2006); Charlie Q.L. Xue and Guanghui Ding, *A History of Design Institutes in China: From Mao to Market.* (New York: Routledge, 2019); Jianfei Zhu, *Architecture of Modern China: A Historical Critique.* (New York: Routledge, 2009). **10** Yu Zhang, *Zhongguo Duiwai Yuanzhu Yanjiu 1950-2010 [Research of China Foreign Aid 1950-2010].* (Beijing: Jiuzhou Press, 2012). **11** Yizhuo Gao et al. "From South China to the Global South: Tropical Architecture in China under the Cold War," *The Journal of Architecture, 27, no. 7-8 (2022): 979-1011.* **12** Charlie Q.L. Xue and Guanghui Ding, *A History of Design Institutes in China: From Mao to Market.* New York: Routledge, 2019.

13 Xuefei Ren, *Building Globalization: Transnational Architecture Production in Urban China.* (Chicago: The University of Chicago Press, 2011); Qui Li Xue, *World Architecture in China.* (Hong Kong: Joint Publishing Ltd. Co., 2010). **14** Charlie Q.L. Xue et al., "Architecture of 'Stadium Diplomacy' – China-Aid Sport Buildings in Africa." *Habitat International* 90 (2019). **15** In 2009 the College of William and Mary established the AidData database, which contains information on 13,427 projects and China's US$843 billion financial commitments from 2000 to 2017 across 165 countries that received Chinese foreign aid. The database provides access to basic project information (project background, amount of assistance, construction period, milestones, construction status, and specific location) listed under the project name and number. **16** Fulong Liu F. and Yu Cao, "Gaige kaifang sishinian: zhongguo duiwai yuanzhu lichen yu zhanwang [40 years of open door and reform: the history and prospect of China's foreign aid]." *Gaige [Reform]* 10 (2018): 52-59. **17** Xia Ai, "18 Years of Overseas Design Practice of CCDI Group as a Complex Journey." In Charlie Xue and Guanghui Ding, eds. *Exporting Chinese Architecture: History, Issues and "One Belt One Road."* (Singapore: Springer, 2022): 229-243.

I.

The Architecture of the Belt and Road Initiative: A New Architectural Order?

Gift Complexes. Hybridizing Extra-State Architecture

The vast Gwadar port is a key logistical route along the new Maritime Silk Road; the buildings close to the port, for example the Gwadar Free Zone Business Center and the Pak-China Technical and Vocational Training Institute, emerge as captivating architectural pastiches bearing witness to the collaborative endeavors between China and Pakistan. The buildings in question feature courtyards and gardens inspired by traditional Chinese architecture, coupled with Arabic-style pointed arches and lattice ornamentations created using cutting-edge techniques; their design combines contemporary and traditional styles and international and vernacular components, creating a paradoxical yet harmonious feature developed within the framework of the development cooperation established between the two countries. Much like the Gwadar Free Zone Business Center and the Pak-China Technical and Vocational Training Institute (see *Pakistan-China Technical and Vocational Institute, Gwadar, Pakistan*, p. 116), many buildings situated along pivotal infrastructure routes do not directly serve the global infrastructure development nor do they fulfill specific economic functions. Instead, they stand as incidental byproducts of business activities intersecting with the needs of local communities. Leisure, commercial, and medical spaces often emerge as indirect interventions within the geographies of the Belt and Road Initiative (BRI).

Despite the seemingly diverse nature of these buildings scattered across continents and countries, for example the Business Center in Gwadar, the hospital in Niger (see *General Hospital of Niger, Niamey, Niger*, p. 110), and the Sino-Italian Cultural Exchange City Reception Center in the Chinese province of Sichuan (see *Sino-Italian Cultural Exchange City Reception Center, Chengdu, China*, p. 104), they do exhibit several similarities. Largely funded by the Chinese government without anticipating reciprocal returns on investments,[1] these buildings serve similar political purposes within the framework of cooperative development assistance. However, what remains to be explored is how, in design terms, these building express their distinct identities through their reinvented architectural languages. How do these structures employ different media-oriented architectural styles – both eccentric and eclectic or sober and modernist – in order to divert attention from what local residents and the global audience consider to be the BRI's contentious political, financial, and social aspects? Frequently gifted from one state to another as donations, these buildings provide us with the means to explore the contemporary fusion of local and global architectural languages, bridging the gap between distant and nearby regions and unraveling the intricate transnational complexities inherent in the BRI's geopolitical landscape.

Indeed, these buildings' apparent lack of immediate economic purpose is precisely what makes them profoundly symbolic. Recalling Venturi's ideas,[2] they symbolize the ordinary, encapsulating the essence of both the present and the past, and personifying the significance embodied by architecture. Amid the prevalent repetitive, mechanical, and utilitarian structures characterizing the urban development that is part of global infrastructure, these symbolic buildings stand out as eclectic urban materials – fragments brimming with symbolic power, serving as vehicles for cross-cultural expression through the art of architecture.

Transcultural Practices and Contact Zones

Exploring cross-cultural exchanges within the field of architecture is not a new undertaking; however, this topic has recently attracted greater attention, fueled by the rapid global dissemination of architectural expertise. In the last two decades, studies focusing on "nomadic expertise" and "global experts" have increased and expanded the historiography and theory of contemporary architecture.[3] Closer examination of contemporary architectural practices within the expansive realm of global infrastructure development reveals the undeniable presence of transculturalism as a prevailing condition of contemporary architecture. Nonetheless, this condition demands a more profound commitment and comprehension, one that surpasses the conventional rhetoric of globalization or colonization that has characterized the academic debate for so long. Indeed, buildings "donated" from one nation to another within the framework of power dynamics have traditionally been chiefly conceived as architectural exports or manifestations of asymmetrical globalization in post-colonial contexts. However, in the past few decades scholars from diverse fields have forged a new path aimed at transcending the concept of binary dichotomies; they have espoused comprehensive transfer concepts such as "translation," "exchanges," "contact zones," and "reciprocal comparison" in order to more comprehensively fathom the intricate nature of cross-cultural phenomena.

Of all these ideas, the conceptual lens that probably uncovers the most insightful interpretation of the transcultural phenomenon in architecture is that of the "contact zone," initially espoused by the renowned scholar Mary Louise Pratt. She defined "contact zones" as social spaces where transculturalism takes place and cultures meet, clash, negotiate, and struggle for power.[4] The contact zone concept was embraced in the field of architecture by Tom Avermaete. Throughout his scholarly pursuits he skillfully harnessed the contact zone concept to unveil the intricate fabric of knowledge transfers and exchanges in Cold War architecture during development programs.[5] He posits that these architectural projects play a pivotal role in facilitating encounters between cultures with diverse fields of knowledge and interests, thereby leaving an indelible imprint upon architectural design and planning. Avermaete emphasizes that within these contact zones, new buildings bear no resemblance to mere replicas of Western architectural models, but rather emanate as manifestations of their own distinct and inherent logic.[6] However, while contact zones foster an environment enabling interaction and the exchange of ideas, it is imperative to acknowledge their latent potential to also engender conflicts and contentions. In architecture, these conflicts have to be seen as generative forces. Indeed, as highlighted by Jorge Mejía Hernández and Cathelijne Nuijsink, innovative and critical approaches to architecture very often emerge during discussions and the resolution of disputes thanks to the exchange of knowledge and expertise.[7]

The institutional framework of the Pakistan-China Technical and Vocational Institute under the BRI (see *Pakistan-China Technical and Vocational Institute, Gwadar, Pakistan*, p. 116) is an outstanding example of these transnational processes.[8] In this respect, it is crucial to first and foremost acknowledge Pakistan to be one of China's fundamental partner countries in the grand plan of the BRI; within this framework the China-Pakistan Economic Corridor (CPEC) commands a prominent position as a flagship project, boasting six sprawling economic corridors alluding to the advent of the New Eurasian Land Bridge. The CPEC framework is decided by the Joint Cooperation Committee (JCC), a collaborative team annually co-chaired by delegates from Pakistan's Ministry of Planning Commission and China's National Development and Reform Commission. This is the context in which the ambitious endeavor of the Gwadar Port project is taking place; it involves a whole host of architectural projects funded, designed, and built thanks to the economic and technical resources of Chinese enterprises. Each project features a different technical task group made up of multidisciplinary experts. It is here that the contact zone initiates dialogue in the architectural exchanges of knowledge at different levels; the latter include different interests, movement of ideas, technologies, construction methods, and variations in the context. According to Mary Louise Pratt, in this environment cross-cultural knowledge and ideas intertwine and lead to a convergence of both countries' architectural cultures.[9]

More specifically, the architectural design proposals for the center were formulated under the guidance of the China Communications Construction Company (CCCC), but they were still subjected to reviews and collaborative exchanges by both a local and Chinese technical task group. The Pakistani team, made up of project managers and a group of multidisciplinary experts, acted as the sentinels of this critical assessment. Similarly, at the other end of the spectrum, the technical working group from China assembled a resolute group that included representatives from China's Ministry of Commerce (MOFCOM), the Chinese embassy in Pakistan, architects, engineers, economists, and project managers, all coordinated by the CCCC. Within the intricate fabric of this expansive endeavor, the interplay between these two groups led to the creation of a contact zone that witnessed the flourishing of transdisciplinary cross-cultural collaborations and negotiations, focusing on every aspect of the project that required attention.

In this contact zone, multiple design proposals for the Gwadar Business Center encountered a series of difficulties and deviations when the local technical task group and many other influential actors scrutinized the entire project. The initial design proposed by the CCCC was a modern box with Chinese characteristics. However, the Pakistani technical task group urged there be changes, in particular several alterations that would include references to the extensive architectural heritage of Gwadar, a mixed legacy influenced by styles adopted in Arabia and Persia. And in fact the design proposal was drastically revised. The building's envelope took the form of a U-shaped layout; this bold vision embraced Islamic architectural elements, for example Arabic lattice ornamentations and pointed and rounded arches inspired by Gwadar's vernacular mud architecture. As regards the program, the final design of the Pakistan-China Technical and Vocational Institute combines the region's distinctive architectural features with the pragmatic needs of its primary users who come from China. In response to their specific requirements, the interior spaces of the building hosts Chinese restaurants, apartments, and secure basement bunkers, ready to serve as shelters in case of an emergency. At the same time, the needs of the local community were duly acknowledged and provisions were made for halal restaurants and mosques, thus creating a culturally integrated space.

Negotiations took place within the contact zone regarding the spatial design and construction of the building. Given the lack of a proper modern construction industry in the city of Gwadar, CCCC opted to import most of buildings, components from China, including electrical equipment, appliances, windows, and decorations, as well as prefabricated structural concrete elements; the latter were then assembled in Gwadar. All these elements were employed using high-tech construction techniques that respected the most advanced Chinese standards and the center's functional requirements. However, one mandatory premise was that the Chinese design standard could not be in contradiction with local codes. For this reason, CCCC repeatedly met with the Pakistan technical task group to discuss all possible inconsistencies and ensure that Chinese standards were appropriate according to the prerequisites of the Pakistani codes.

Furthermore, although the project managers and engineers came from the CCCC, the labor force was made up of a team of Chinese and Pakistani workers who were trained by Chinese engineers to assist them during construction. During this process the laborers used WeChat, seamlessly scanning the QR codes on each building's components; this allowed them to undertake multiple tasks at the same time without having to receive explicit instructions. In addition, the local workforce's tasks also included informing the Chinese engineers about local climate conditions and other factors that might affect the construction of the building; this led to a reciprocal exchange of information. In the words of Mary Louise Pratt, this contact zone elevates both participants to a newfound state of normalcy, transcending cultural boundaries and generating a space of shared understanding.[10]

In this context, the contact zone concept emerges as an innovative methodological instrument, boosting our capacity to achieve a deep-rooted understanding of the global practice behind architecture production. This approach saw the involvement of several actors, aid agencies, and construction companies as well as combined political motivations, interests, and legal systems in the transfer of architectural models, thereby providing considerations regarding transnational modes of transfer from the West to non-Western and multidimensional globalization.

Hybrid Languages and Transnational Identities

The dynamic interplay of transcultural operations within the contact zones is manifest as hidden substructures, but more so as linguistic expressions fueling contemporary architectural projects. As a result of exchanges, collaborations, conflicts, and negotiations, the aforementioned Pakistan-China Technical and Vocational Institute displayed, for example, an intricate hybridization of architectural elements and spatial typologies typical of China and Pakistan; as a result, these elements and typologies become a distinctive feature of the architecture built in a development aid context.

However, the Pakistan-China Technical and Vocational Institute is not the only exemplar of architectural hybridization within the Belt and Road Initiative (BRI) where transcultural operations are at play. Another example is the Djibouti National Library and Archive Center showcasing a transcultural architectural scheme that incorporates elements of Chinese design philosophy, e.g., symmetrical courtyards, but embellishes it with recognizable Islamic architectural features, e.g., a series of centrally placed pointed arches juxtaposed against Moorish arches at the corners, all presented in a modern style. As a global phenomenon characterized by unpredictable interchanges, the theme of the architectural pastiche within the BRI can be even more varied; one example would be the Sino-Italian Cultural Exchange City Reception Center in Chengdu. In this case the spatial devices and architectural elements imported from China and Italy are combined, generating a rich tapestry of juxtaposition and overlap. The project's dual sections, separated by water and flourishing greenery, respectively symbolize the distinctive features of Italian classical architecture and Chinese architectural heritage. On the one hand, circular and semicircular walls and windows, spiral stairs, and rounded fountains echo the Renaissance theme of circular perfection, on the other, elegant gable roofs, inner courtyards, and circular openings are reminiscent of the classical gardens of Suzhou, a city nestling in a peaceful bamboo forest; these features reflect the relationship between humans and nature, as emphasized by Feng Shui philosophy. Here, the space is eclectic, the façade is eclectic, and the landscaping is also eclectic; the two styles are linked by ubiquitous spaces carefully and deliberately organized to express cultural meanings (see *Sino-Italian Cultural Exchange City Reception Center, Chengdu*, China, p. 104),

The architectural styles of these buildings represent the respect accorded to the local *pastiche*. The architectural strategy is to balance exogenous and endogenous architectural features in a modern manner, but with a focus on native features. These architectural endeavors, full of symbolic forms and languages, take their cue from the direct extension of multiple vernacular architectural elements. The architecture becomes almost an explicit text which, however, lacks any critical understanding of regional traditions and their reinterpretation in the contemporary setting. In this sense we cannot agree more with Thorsten Botz-Bornstein when he states that this is not an outrageous aspect of this architecture.[11] These buildings show that regionalism does not have to be critical. Indeed, as further indicated by Botz-Bornstein, the notion of a self-critical movement, such as Critical Regionalism, is intrinsically linked to the Western tradition of enlightenment – a reality that can be both advantageous and problematic, especially when these endeavors are introduced in contexts where the Western tradition of critical thought does not hold sway or may even be nonexistent.[12]

The astonishing mix of stylistic elements, deliberately researched by the design teams within the BRI, clearly illustrates the prevalence of an iconic conception of the architectural and spatial devices, in which the symbolic meaning of each element is mainly derived from, and an expression of, cultural issues. This kind of architecture is not just an end in itself, but a means used for political and financial purposes. The work performed by Lawrence Vale has already highlighted the political use of symbolism in architecture, a mechanism deeply rooted in this field.[12] However, unlike his capitol buildings that bear a national identity, within the framework of global infrastructures, such as the BRI, the identity is transnational and thus the language must be hybrid. In the case of architecture which is, in a sense, "gifted," the act of conveying meaning becomes paramount since its goal is to represent the aspirations of stakeholders from different countries. This process is embedded in the broader mechanism of the BRI, as it is in the construction of other global infrastructure, thus becoming deeply rooted in architectural engagement. Identities intertwine, often challenging easy recognition, with the building's ability to evoke sensation thus assuming paramount importance. As a result of ongoing negotiations between the parties involved, the architecture embraces a hybrid, relinquishing purity and coherence in favor of inclusiveness. Embracing different cultural issues and standards of practice becomes an important characteristic worthy of being investigated. No single entity prevails; no longer colonizer or colonized, but rather mutual cooperation achieved and turned into architectural terms thanks to symbolic and cultural issues. One could say that within the Belt and Road Initiative (a framework brimming with abstract ideas, concepts, ideologies, and rhetoric) perhaps the most tangible aspect of development aid lies in the transcultural dimension of its architectural projects.

Architecture: Form and Meaning

As briefly demonstrated above, the engagement in transcultural practices within the context of global infrastructure amplifies the mobilization of the stakeholders involved in the production of the built environment. In terms of the design process and final outcomes, all negotiations, adaptations, and exchanges must assume new values, ones which surpass the notions of imposition, purity, and clarity that have long dominated our understanding of good architecture in the Western world. This shift entails a social and institutional process that revives eclectic architecture as an alternative to formal coherence and conceptual precision.

Within architectural discourse, where style and substance, meaning and form, language and function, have traditionally been seen as opposing forces, the proliferation of "gift complexes" in contemporary settings offers a different perspective. It anticipates a more fluid comprehension wherein pragmatic reasoning imbues ornamentation with functional and political purpose, transforming it into the essence of architectural expression and construction. The concepts of "contact zones" and "transculturalism" further help us to recognize that the understanding of architecture is intricately interlinked with a broad process of exchange, formation, and coordination involving the stakeholders, from different countries, responsible for the design and construction of contemporary architectural artifacts.

In a manner reminiscent of 19th-century tradition, the new eclecticism of transnational architecture revives the notion of "architecture that tells stories." Meaning takes precedence over coherence, as symbolism does over spatial effect or the overall conceptual narrative. It signifies a return to language and a yearning for architecture to communicate; as observed by Manfredo Tafuri, this is an inherent and never-ending quality.[14] However, unlike Las Vegas, as described in the words of Venturi, Izenour, and Scott Brown,[15] this is not only an architecture of bold communication, but also an architecture of evocation, reminiscence, and dissemination thanks to its explicit architectural artifacts that evoke imagery and sensation rather than just messages.

Indeed, as suggested by Roland Barthes, within the realm of cultural aesthetics, Oriental aesthetics diverge significantly from the Western paradigm that emphasizes fixed and univocal interpretations of signs and symbols.[16] Oriental culture instead embraces a more malleable, unconstrained approach, receptive to multifarious interpretations of these signs and symbols. In this cultural milieu, knowledge is found in the allure of ambiguity, revealing the complexity of the layers of meaning embodied by signs and symbols. It recognizes the weighty significance of intuition, subtlety, and suggestion as vehicles to impart meaning. Barthes astutely underscores that the Oriental approach to meaning-making epitomizes an immersive entanglement with the sensory and aesthetic tapestries of existence, exalting the ephemeral, the evanescent, and the poetic, and often evoking an air of mystique and contemplation.[17] In this sense, "gift complexes" reveals that the means of architectural communication are multiple and diverse. Yet they can also appear superficial; these hybrids of strong and soft symbols can represent the essence of architectural conception within the intricate planning and design processes.

These issues force us to reconsider the juxtapositions that traditionally define boundaries between exterior and interior, envelope and space, form and sign. Instead, when all these elements are taken collectively they can become communication tools, reflecting the various ways in which we interpret buildings, not only as experts, but also through the unconscious associations triggered by our cultural backgrounds and knowledge of the world. In other words, "gift architectures" allow us to reconsider the relationship between form and meaning in architecture in order to avoid reducing it to a simplistic binary opposition, but instead expand it to encompass a more complex theoretical and analytical description which, by organizing geometrical characteristics and spatial configuration, is used in architecture to build a relationship between conceived and perceived aspects of space.

1 While more generally, buildings and constructions within the framework of the BRI feature an institutional scheme where both the recipient country and China equally contribute to financing their realization, artifacts in the gift complexes category area are characterized by unilateral investments from China as a means of diplomatic development with recipient countries. **2** Robert Venturi, *Complexity and Contradiction in Architecture*. (New York: The Museum of Modern Art, 1977). **3** Donald McNeill, "Skyscraper Geography." *Progress in Human Geography* 29, no. 1 (2005): 41–55. **4** Mary Louise Pratt, "Arts of the Contact Zone." *Profession* (1991): 33–40. **5** Tom Avermaete, "Coda: The Reflexivity of Cold War Architectural Modernism." *The Journal of Architecture* 17, no. 3 (2012): 475–477. **6** Ibid.

7 Jorge Mejía Hernández and Cathelijne Nuijsink, "Architecture as Exchange: Framing the Architecture Competition as Contact Zone." *Footprint* 14/1, no. 26 (2020): 1–6. **8** All the information and concepts regarding the Pakistan-China Technical and Vocational Institute have been developed in Sohrab Ahmed Marri, *Architecture for "Other." China's New Eclectic and Pragmatism in Developing Countries within the Framework of the Belt and Road Initiative.* (Doctoral Dissertation, 2021). **9** Mary Louise Pratt, "Arts of the Contact Zone." *Profession* (1991): 33–40. **10** Ibid. **11** Thorsten Botz-Bornstein, *Transcultural Architecture: The Limits and Opportunities of Critical Regionalism.* (New York: Routledge, 2017). **12** Ibid. **13** Lawrence J. Vale, *Architecture, Power, and National Identity.* (New York: Routledge, [1992] 2008). **14** Manfredo Tafuri, *Progetto e utopia: architettura e sviluppo capitalistico.* (Rima: Laterza, 1973). **15** Robert Venturi, Denise Scott Brown, and Steven Izenour. *Learning from Las Vegas.* (Cambridge: The MIT Press, 1972). **16** Roland Barthes, *Empire of Signs.* (New York: Macmillan, 1982). **17** Ibid.

Worlds of Special Rules. Architecture between Humans and Data-Driven Machines

The Free Zone of Khorghos or Horgos is located near the farthest point on earth from any ocean (also known as the "Eurasian Pole of Inaccessibility"). Half of the Free Zone (called Horgos) is situated in China, and half in Kazakhstan (known as Khorghos): it is an important example of how contemporary production and logistics are turned into spatial forms. Here, the stage is set for a complex interplay of regulations, technologies, and spaces, where the flow of commodities and people intersect with the architectural organization of checkpoints, customs zones, warehouses, manufacturing facilities, and trading areas. Millions of different goods transit at this 'gate' every day, a constant flow subjected to the meticulous scrutiny of an electronic checkpoint system that acts as the entryway to the International Center for Boundary Cooperation (ICBC) in Khorgos. This controlled route marks the beginning of multiple sequences, activating not only the movement of goods, but also their temporary storage in the customs control zone, stocking, and trading areas. An intricate sequence of movements unfolds – a process where an automated logistics management system sets the rhythm of the duration of the storage, restricted by a strict three-hour limit, and automated vehicles establish the spatial arrangement that enables the efficient actions of shipping containers. Yet, amid this strongly automated activity, human agency persists. The carriers, notified of the arrival of the goods, engage in a controlled exchange with official personnel, before moving the goods and raw materials to the manufacturing hub. At the same time, thousands of visitors of different origin cross the border every day to shop for lower-priced Chinese goods. This logistical system extends beyond the boundaries of the ICBC checkpoint. Customs inspections, controls, trade, and accommodation are present within an expansive territory of 5.28 square kilometers, still under further development, where different kinds of architectural artifacts become material supports in the act of production and exchange: trade centers, temporary storage spaces, shops, manufacturing facilities and exhibition centers, to name but a few, generate a backdrop against which industrial structures become part of the intertwined narrative of rigid controls and free movements characterizing the site. As the implementation of global logistics brings production ventures to the farthest corners of the globe, the new factory finds itself no longer isolated. New automated production lines, such as the one manufacturing service robots for Boshihao Electronics, are established in the Khorgos Special Economic Zone and superimposed on a vast territory intended to host their supporting facilities which include not only logistics activities and basic infrastructure, but also international universities, hotels, sports complexes, and a trading center (see *Khorgos Special Economic Zone, Khorgos, Kazakhstan*, p. 142).

Such exchanges epitomize the essence of this space, where the world of automation is intertwined with the human presence: this short narrative provides an opportunity to contemplate the notion of logistics and trade through an architectural lens, or perhaps even as issues of spatial consideration. Free zones and hubs of contemporary production, with their astonishing, avant-garde logistical systems, transform the canvas of the Belt and Road Initiative into a mechanized panorama. Here, the ebbs and flows of individuals and commodities become instrumental in molding the built landscape. These intricate connections between manufacturing, logistics, research, and even leisure, engender architectural hybrids and an intricate urban fabric. They reveal the modern industrial enclave as more than merely a utilitarian space, but as an infrastructural realm that materializes through protocols and practices stipulated by the global capitalist order.

Automated Practices and Machinic Landscapes

Automation, once confined to the realm of production, has rapidly metamorphosed into a spatial issue that resonates in the practices and discourses of architects and urban planners. This shift echoes the broader transformation of our society as it grapples with the permeation of automation into the very fabric of our lives. Architecture is no longer a passive recipient of technology; it has become a proactive medium through which automated practices, involving both human and non-human agency, find their expression. From buildings that adjust their environment based on real-time data to sensor-integrated urban landscapes, the fusion of architecture and automation has already ushered in a new era of spatial possibilities.

However, the BRI, with its expansive global network of logistical hubs and information systems, exemplifies another interesting marriage of automation and architecture. It displays how the design of free-trade zones and industrial parks, which nowadays can be considered an important part of the DNA of contemporary urbanization, is no longer confined to the materiality of concrete columns, steel structures, or brick walls; it encompasses, and is driven by, the minute calculation of machines, sensors, and algorithms that orchestrate an intricate arrangement of goods, services, and information according to the logistics regime. Keller Easterling states that "some of the most radical changes to the globalizing world are being written in the language of this almost infrastructural spatial matrix."[1] This conceptual shift compels us to view architecture not merely as static edifices, but as a living, responsive entity that engages with the flows of data, people, and commodities. As we delve deeper into the architectural manifestations of the BRI, we encounter these spaces in which the material structure and the immaterial infrastructure collaborate in the construction of a machinic landscape that transcends local boundaries and involves both distant networks and closer articulations.

The Lianglu-Cuntan Free-Trade Port Area in Chongqing, China, stands as a compelling testament to the important role of automation and logistics in contemporary architectural discourse. Envisioned at the confluence of the New International Land-Sea Trade Corridor and the Yangtze River Economic Belt, this extensive logistical node epitomizes the essence of the Belt and Road Initiative due to its intricate integration of large-scale infrastructure and new built forms. Spanning 3.88 square kilometers, and inaugurated in 2020, the port's profound influence resonates as a logistical hub serving both regional and global networks. The whole area, dedicated to port, stocking, and manufacturing functions, transcends traditional notions of spatial demarcation. Its architectural fabric – little shelters, prefabricated boxes, and big warehouses – merges seamlessly with a complex mechanical system made of multi-story concrete piers, large steel cranes, railing systems, and X-ray controlling machines, all driven by complex management software programs. They generate a mega-structural landscape blurring the boundaries between the built environment and automated machines. This multimodal infrastructure serves as a platform to orchestrate the hierarchical movement of containers based on their destinations, creating a spatial organization that mirrors the complexity of logistical operations. Several infrastructural nodes (e.g., the implementation of a state-of-the-art equipment CT-type H986 X-ray inspection machine for customs clearance, capable of examining 200 12-meter-long containers per hour through a system of artificial intelligence) exemplify how in these spaces automation optimizes organizational and spatial efficiency (see *Lianglu-Cuntan Free-Trade Port Area, Chongqing, China*, p. 130).

Everything is sized and controlled to enhance the efficiency of the machine over human agency, which has almost no control over the space anymore. The entire landscape, in fact, is meticulously organized around the standardized dimensions of shipping containers and physical mechanisms, such as loading docks, radiofrequency identification (RFID) technology, global positioning system (GPS) tags guiding automated robot vehicles and other modular infrastructures and governing their movement and interlocking. From harbor depths, crane spans, and storage sites to tunnel heights and turning radii, every architectural detail is designed to conform to the shipping container's dimensions and the flows of global trade it represents. In this regard, Aaron Tobey, while discussing the new transportation landscape in the Port of Tokyo, has used the concept of "quantified space" to describe the spatial and structural set that allows shipping containers to operate as abstract units organized by a set of standardized physical infrastructures and immaterial protocols.[2] The end result is a system where any container can traverse any route, be handled by any port, transloaded onto any truck or railcar, and reach any destination – a testament to the power of standardization in a globalized world.[3]

Digital information finds its main architectural expression within this realm. Flows of data and information underpin the interconnectedness of logistics and infrastructure while shaping spatial organizations. Automated systems, incorporating real-time data analysis, facilitate swift decision-making and heightened operational efficiency. This integration transforms the built environment into a dynamic repository of data, reshaping how human and non-human occupants interact with and shape architectural spaces. In this sense, contemporary architectural infrastructure is no longer solely material or digital, but an intertwined co-presence of these two aspects of reality that both contribute to the spatial formation of a new machinic landscape. Yet, as Jesse LeCavalier observes, "Architecture has always been a machinic landscape." Our focus should now be directed toward a deeper understanding and regeneration of existing infrastructure, to the invention of new hybrid typologies and, more broadly, to the fostering of spatial dispositions that recognize the political implications of logistics as inherently architectural.[4]

Mixed Flows and Metabolic Landscapes

The notion of flows – the seamless movement of goods, people, and information – lies at the heart of automation, logistics, and procedural activities in architecture. As previously illustrated, this interconnectedness requires robust infrastructural frameworks that facilitate the spatial arrangement, synchronization, and movement of these flows; the machinic landscape is the result of a broader metabolic landscape, which nowadays defies local and territorial boundaries.

As observed by Daniel Ibañez and Nikos Katsikis, the notion of "urban metabolism" as a driver of spatial strategies is not new. From Patrick Geddes's Valley Section to the monumental structures envisioned by the Japanese Metabolists, the idea of a city and an architecture organically intertwined with the movements of humans, materials, and energy has woven its way through the mid-20th century.[5] Yet this concept, historically confined to a regional dimension, comes face to face with an era characterized by an unprecedented complexity and a planetary-scale amplification of metabolic relations. As further noted by Ibañez and Katsikis, contemporary discussions of metabolism often stumble at a crucial juncture: the integration of formal, spatial, and material attributes. Positivistic approaches related to the technological sphere often lean toward performative interpretations of flows, while theoretical explorations of the sociopolitical dimensions of metabolic processes tend to sidestep their formal spatial ramifications. In this milieu, the realm of architectural design neglects the sometimes unyielding qualitative and spatial nature of landscapes and infrastructures.[6]

While examining some of the most recent global infrastructural spaces, all these ideas of flows and their spatial arrangement have a rather particular repercussion in the architectural realm. Flows not only shape the distribution of functions and the arrangement of architectural elements in space and time, they also determine their dimensions and characteristics. The materialization of space in this context serves to accommodate not a single entity, but rather a multitude of diverse flows – humans, vehicles, cargo, goods, robots, information – each meticulously arranged to enhance efficiency and adaptability at various levels and scales. In this domain, as insightfully pointed out by Liam Young, we find ourselves in the middle of a transition toward a "Post Anthropocentric architecture, in which the human body is no longer the dominant measure of space."[7] In this case it is the symbiosis of human and machine that defines the parameters, dimensions, distribution, and features of architecture. While this paradigm shift is not entirely unprecedented, it assumes an exaggerated form within logistical hubs, ports, warehouses, laboratories, and factories, where the human presence is at times limited or even restricted. This is, for instance, the case of the Prologis Logistics Center in the new area of Liangjiang in Chongqing: in the four, two-floor warehouses, the organization of the buildings is driven by the ramps that allow trucks to move three-dimensionally both inside and outside it (see *Prologis Logistics Center, Chongqing, China*, p. 136).

According to Negar Sanaan Bensi and Francesco Marullo, despite the high level of technical automation, logistics and manufacturing does not only produce generic spatial configurations – like fulfillment centers, container terminals, or interchange yards – they also create specific forms of employment and new spaces to host this workforce.[8] New urban zones emerge as part of the program of the BRI, for example the Great Stone Industrial Park, an intensive manufacturing hub situated in the vicinity of Belarus's capital, Minsk, or the China-Egypt TEDA Suez Economic and Trade Cooperation Zone, a large free zone located a few kilometers from the Suez Canal. These new urban zones encompass hybrid urban structures that accommodate a varied program ranging from automated production spaces to places for leisure and consumption: these zones include a series of supporting amenities, such as housing, offices, hotels, conference centers, and even an amusement park. All these elements are organized within a big grid pattern generating large plots measuring 300–500 meters by 300–500 meters; a multi-lane road network connects the different functional areas; the means of transportation include cars, trucks for cargo, and even dedicated buses for workers and their families. Although the flows of people, goods, and machines might seem similar, they feature distinct circulatory systems, given the different natures of their respective loads and transportation means – animate versus inanimate. This new urban fabric generates a series of interstitial spaces that are doing more than just supporting mechanical movement and circulation; the presence of human flows means they essentially act as "public spaces" with direct reference to their everyday users[9] (see *Great Stone Industrial Park, Minsk, Belarus*, p. 148, and *Suez Economic and Trade Cooperation Zone*, Suez, Egypt, p. 154).

Architecture: Humans and Machines

In our ever-evolving architectural landscape, characterized by the rapid rise of automation and the sprawling reach of global infrastructures, the intertwining of human presence and machine-driven processes has become an essential facet of the architectural practice. Indeed, acknowledging this fact is a significant step toward admitting that the distinction between human and non-human is disappearing, a notion explicated by Bruno Latour in his concept of "Flat Ontology." Latour invites us to discard conventional hierarchical views that have long privileged human actors above their non-human counterparts. Instead, he asks us to perceive all entities – be they human or non-human – as equal in terms of agency.[10] This is a departure from the norm, urging us to challenge a positivistic understanding of rhetorically named human-centric design practices. Indeed, it helps us to reconsider how designing with non-humans in mind pushes us to consider the profound impact of our creations on not just the human experience, but also on the environment at large. It forces us to recognize that the architecture we design inevitably influences our landscapes, our use of resources, and the very essence of our existence as humans.

For this reason, a broader understanding of architecture within global infrastructure should prompt us to pay more attention to those places, such as the industrial parks, free zones, and ports which, despite the extremely important role they play in shaping the contemporary urban phenomenon, have been largely overlooked in the general discourse on architecture. According to Pierre Bélanger, these spaces are more the product of engineering than meticulous design or planning.[11] And yet, if we abandon the dichotomous and hierarchical distinctions between human and non-human actors within the built environment, we could open the door to a reimagining of these spaces as valid subjects of architectural exploration and innovation. This journey of inquiry could not only offer a fresh critique of architecture's often anonymous and speculative practices, but also underscore the fundamental role played by architectural design in shaping the world at large. It is a collaborative venture that seeks to elevate the ostensibly mundane aspects of industrial and logistical infrastructure within architectural discourse, raising awareness among stakeholders about their pivotal role in the ongoing transformation of urban life.

Moreover, as we advocate for this paradigm shift, the developments along the BRI exemplify how automation also blurs the boundaries between digital and material realms of space. This gives further resonance, in a renovated manner, to Manuel Castells's concept of a "space of flows," where digital networks and swift transportation corridors merge to facilitate time-sharing social practices.[12] This idea breathes new life into our understanding of space, proving that it is a dynamic entity inextricably intertwined with time. Castells's notion of the "space of flows" emerges as a highly relevant conceptualization within contemporary spatial theory, emphasizing the crucial significance of logistical territories as its material foundation.[13] Such a perspective challenges traditional distinctions between static and dynamic elements in contemporary architecture, revealing just another aspect defying established categorization. In this sense, we must acknowledge that a profound shift is taking place in our understanding of flows and their embodiment in space. New forms of spatial flexibility and distribution are emerging, reshaping our perception of control and spatial constraints. The age-old binary of spatial flexibility versus structural rigidity crumbles away. As our spaces embrace the confluence of human and non-human agents, driven by material and immaterial forces, new conceptual frameworks are required to interpret architectural opportunities within the context of global infrastructure development.

1 Keller Easterling, *Extrastatecraft: The Power of Infrastructure Space*. (New York: Verso, 2014). **2** Aaron Tobey, "Architecture at Sea: Shipping Containers, Capitalism and Imaginations of Space." *Architecture and Culture* 5, no. 2 (2017): 191–212. **3** Ibid. **4** Jesse LeCavalier, "Human Exclusion Zones: Logistics and New Machine Landscapes." *Architectural Design* 89, no. (2019): 48–55. **5** Daniel Ibañez and Nikos Katsikis. *New Geographies 06: Grounding Metabolism*. (Cambridge, MA: Harvard University Press, 2014). **6** Ibid. **7** Liam Young, "Neo-machines. Architecture Without People." *Architectural Design* 89, no. 1 (2019): 6–13. **8** Negar Sanaan Bensi and Francesco Marullo, "The Architecture of Logistics." *Footprint* 10 (2018): 1-8. **9** Ibid. **10** Bruno Latour, *Pandora's Hope: Essays on the Reality of Science Studies*. (Cambridge, MA: Harvard University Press, 1999). **11** Pierre Bélanger, *Landscape as Infrastructure: A Base Primer*. (New York: Routledge, 2009). **12** Manuel Castells, *The Informational City*. (Oxford: John Wiley & Sons, 1991). **13** Negar Sanaan Bensi and Francesco Marullo, "The Architecture of Logistics." *Footprint* 10 (2018): 1-8.

Mass Housing Enclaves. Between Standard Forms and Local Conditions

The brand new residential development built in a desert area a few kilometers south of Luanda, the bustling capital of Angola, features repetitive rows of buildings (a total of 750), each with seemingly few architectural variations. This colossal endeavor – one of the largest housing projects built by a Chinese firm on foreign soil – is known as Kilamba Kiaxi, or Kilamba New City. It is just one of the many housing projects being built in various parts of the world and features an evident repetition of architectural elements, both in this specific site and in relation to global patterns of similar developments (see *Kilamba Kiaxi, Luanda, Angola*, p. 186).

The Belt and Road Initiative (BRI) passes through different and occasionally extreme landscapes, ranging from deserts to man-made islands. Within this framework, navigating housing geographies proves to be a formidable undertaking. Against this diverse backdrop, the living spaces along the routes of the BRI primarily take the form of expansive compounds, where discernible distinctions between one location and another blur when scrutinized through the lenses of architectural morphology, functional programming, and construction systems. Nevertheless, it is precisely this architectural feature that is significant. It reveals the essence of an urbanizing world, where mass-produced, medium- to high-rise buildings not only provide accommodation for a substantial segment of the global urban population, but also offer the aspirational dwellings envisioned by many.

While it is impossible to deny the allure of iconic images portraying endless series of replicated buildings in the most disparate corners of the developing world, the apparent image of universality and homogeneity reflected by these structures and urban settings requires closer examination. Indeed, as pointed out to us by urban scholar Davide Ponzini, the overarching rhetoric of globalization that renders places uniform and unvaried in pivotal hubs of contemporary interconnectedness not only oversimplifies the intricate socio-spatial nature of these sites, but also hampers a closer interpretation of the intricate dynamics that coexist between the forces of homogenization and the spaces of uniqueness, thus risking impairing a critical issue: their potential improvement by local and transnational actors.[1] Can the BRI be an effective lens to better investigate these perspectives?

Standard Practices and Technical Adjustments

More than five decades ago, Christopher Alexander highlighted the fact that the bulk of housing production relies on mass production methods, i.e., entailing the construction of numerous houses using semi-automated processes.[2] This practice continues to persist even in our contemporary age; it is particularly evident in the surge of novel developments catalyzed by global infrastructure initiatives such as the BRI. This persistent reliance on standardization and automation has emerged as a recurring motif in architectural discourse, further amplified by the proliferation of digital technologies and the rise in transnational architectural practices.

When examining the repetitive architecture present within the expansive context of the BRI, Keller Easterling's notion of architecture as an "organizational expression of spatial arrangements" emerges as a transformative lens.[3] This conceptual framework shifts the focus away from the mere formal attributes of buildings, channeling it toward the parameters and protocols that intricately shape spatial configurations of the very familiar housing development format. Recent architectural endeavors undertaken by the BRI (e.g., a market-driven housing compound in Georgia, or an affordable housing complex in the Maldives) vividly exemplify this paradigm shift, spotlighting the fact that architecture should now place equal emphasis on buildings' attributes and on a repertoire of sequential operations influenced by companies' dynamics, intricate patterns of connectivity, and the dynamic interplay of multifarious components (e.g., building regulations, ISO standards and local and global real estate market logics). Within the overarching framework of the BRI, all these architectural parameters, protocols, and standards flow seamlessly across global boundaries, adapting to nuanced local conditions; this is achieved thanks to the initiatives that reshape the fabric of the built environment, implemented by state-owned developers, construction companies, and engineering firms. Focusing on these ground-level actors and their contribution to the dissemination of development practices worldwide allows us to first of all understand how an assimilated and internalized assortment of planning and construction standards influences the model of Chinese-led urbanization overseas.

A vivid example illustrating these intricate dynamics is the affordable housing initiative implemented by the China State Construction Engineering Corporation (CSCEC) in Hulhumalé, a man-made island in the Maldives. The project provides new housing for an estimated 30,000 residents in the country's densely inhabited capital. By exploiting a meticulously elaborate business model and operational standards developed during its domestic operations, the CSCEC applied these strategies and operations to an international setting by prioritizing internal resource integration. This transnational exchange required the transfer of deeply ingrained and standardized logics, rules, and practices, together with the consequent migration of labor, machinery, materials, and technologies from China to the Maldives. An on-site management team of Chinese designers and project managers was created during the earliest phase of the project; the specific objective was for it to act as the intermediary between the Chinese design team and on-the-ground realities. The task of the centralized on-site procurement team was initially to help reduce operating costs and enhance efficiency thanks to large volume orders purchased from a dedicated online platform which, in order to comply with Chinese standards, would be imported from abroad. Secondly, the on-site procurement team would also serve to intensify the development of the project by establishing different collaboration agreements with local suppliers and partners. This would ensure that the Chinese contractor had excellent control over its supply chain. Indeed, given the presence of the local procurement team and their codified practices, the CSCEC engineering department required that both general contractors, subcontractors, and all their suppliers had to possess several kinds of qualifications, e.g., the ISO 9001 Quality Management Certification (see *Hiyaa Housing Project, Hulhumalé, Maldives*, p. 168).

As recently noticed by Fernando Lara, in most cases the diffusion of modern standards and technologies has played a leading role in the homogenization of housing solutions worldwide.[4] The use of reinforced concrete slabs, beams, and columns that all meet modern standards has been adapted, for both practical and symbolic reasons, by every construction industry in developing countries. However, even if rarely perceived by external observers, the pivotal impetus behind the transfer and adaptation of technical protocols spearheads new construction techniques and spatial designs. Fueled by the widespread dissemination of organizational knowledge and expertise, the technologies transferred by Chinese companies within the BRI metamorphose to suit specific local conditions. This adaptability is evident in the Maldives where unique environmental factors (e.g., elevated salt levels and humidity) require strategic adjustments of very standard concrete elements and construction technologies developed in China. Construction technologies such as CFG pile composite foundations, prefabricated climbing scaffolding, and aluminum concrete formwork were specially developed by the Chinese contractor to deal with the soil and climate on the man-made island in the Maldives.

The Hulhumalé housing project – a joint endeavor with the local Maldives government – epitomizes the convergence of global standards and local regulations. On the one hand, the project seems to fully respect conventional Chinese standards for distribution schemes, functional zones, and building dimensions. The height of, and distance between, high-rise building blocks – up to 24 floors with a minimum distance of 33 meters between one building and another – are based on China's fireproof and setback regulations. All the residential towers have a central core that acts as a pivot around which standardized apartment units with identical layouts are either mirrored or rotated. Simply designed floor plans – based on the developer's guidelines, codes, and numerical formula in an effort to optimize the layout of high rise buildings – are then extruded vertically along the core without any variations between floors; this keeps production costs and construction time to a minimum.

Then again, the project readily implemented adaptive alterations in order to harmonize with the local normative and cultural context. The towers' central core encloses the building's vertical distribution; it is conceived as an open-ended space, facilitating natural cross ventilation, which is so necessary in the hot and humid climate of the Maldives. The layout of the housing units are carefully planned with a minimum 51 square meter floor area and two bedrooms with en suite bathrooms; this is in order to not only comply with local regulations imposed by the Hulhumalé Planning and Development Organization (HPDO), but also to respect the habits of the inhabitants in the Maldives. During this process the design of the balcony railings and walkways, based on the technologies developed by the CSCEC, were reviewed and modified in order to satisfy the local regulations imposed by the HPDO, i.e., a minimum height of 1.1 meters and a thicker aluminum bar structure. These modifications were internalized by the CSCEC and reiterated in similar developments, thus transforming a specific local constraint into a replicable model which then became a company standard. Although these minor details appear insignificant, they substantiate the statement by Keller Easterling that organizational formats are sometimes improvisational and responsive to circumstantial changes, anomalies, and seemingly illogical contextual forces.[5] Even such minor modifications wield considerable influence as regards iterative spatial production.

Standard Forms and Spatial Imageries

In magnificent renderings and built urban landscapes, Chinese urbanization has frequently materialized as an assortment of skyscrapers, sprawling industrial zones, towering residential complexes, and different codified architectural types, emerging as what Keller Easterling termed "repeated spatial formulas."[6] These dynamic urban configurations are both replicable and transmittable in several contexts; they undergo periodic updates and revisions in order to adapt to needs that change over a period of time. However, these configurations are generated by spatial imageries and, at the same time, contribute to shaping them; in other words, according to Josh Watking, they are representations and discourses which enable, legitimize, and give meaning to collective spatial urban practices.[7] Although these imageries are now coming mainly from the East, they serve as blueprints for brand new urban productions and ways of life worldwide thanks to the protocols and ideals disseminated by China's state-affiliated developers who influence urban aesthetics and use architectural ventures to fuel economic progress.

Within the framework of the BRI, major Chinese real estate developers and construction conglomerates, including companies like China Merchants and CAMC Engineering, cross international borders; they collaborate with their local counterparts in host nations, i.e., political, economic, and cultural stakeholders that serve as the principal conduits for the global dissemination not only of technical protocols, as previously mentioned, but also spatial imageries. A case in point is the Hualing Tbilisi Sea New City, an expansive newly developed 920 square kilometer area superimposed upon the existing fabric of small villages in the Gurjaani region of central Georgia, just outside the capital Tbilisi. As the flagship venture of the Hualing Group (a well-established Chinese real estate developer that injected approximately 500 million US dollars into Georgia), the Hualing Tbilisi Sea New City essentially wipes the slate clean, erasing extant structures to pave the way for novel urban infrastructures that adhere to Chinese spatial ideals of modernity, namely high-rise residential towers, postmodern shopping malls, and large roads adorned with flower beds (see *Hualing Tbilisi Sea New City, Tbilisi, Georgia*, p. 174).

The final vision for Hualing Tbilisi Sea New City encompasses an extensive residential domain, including apartment and villa complexes complemented by a diverse array of amenities. In this desolate landscape the latter include recreational zones, commercial spaces, educational institutions, libraries, exhibition halls, cinemas, public safety facilities, administrative buildings, healthcare centers, and sports facilities. These amenities serve as pivotal marketing instruments for the overall endeavor, offering a taste of urban living and bestowing a bourgeois atmosphere on the environment. The approved urban blueprint mirrors the design of Chinese gated communities with their imposing walls surrounding housing units and dedicated facilities. Likewise, the architecture echoes the aesthetics of Chinese cities, featuring medium-rise embellished structures decorated with very pronounced Occidental façades – faux maroon bricks, ornately framed windows, dozens of little circular balconies, and sprawling landscaping with gracefully curved green islands and water pools. The repeated external architectural forms and embellishments are a meticulous assemblage of Western-style structures and Chinese forms. In the words of Bianca Bosker, this imitation goes beyond architectural and construction techniques.[8] It strives to capture the "essence" and experiential quality of old cities in the West and the Eastern way of living, consolidating a spatial imagery whose aim is to represent a new urban standard of living. Indeed, while this type of gated housing compound is common and well established in most Chinese urban environments, it comes off as something new and novel in Georgia.

The final forms of these residential complexes in global infrastructure development sites reflect a codified practice of Chinese developers who have a clear idea about marketable products; they appear to echo the practices of a real estate sector which, in the fields of architecture and urban governance, tend toward its complete commodification. However, unlike many studies focused on "commodity housing," our attention shifted from housing consumption to the production and promotional practices used by real estate developers to assert that houses should not be regarded as mere commodities, essentially indistinguishable one from another. Instead, the differentiation achieved through the company's standardization of architectural elements is strategically employed to cater to a diverse array of cultural practices and urban lifestyles which, as previously noted by Sharon Zukin, are multiple, diverse, and relative to cultural contexts.[9] This kind of company standardization must be perceived as a way for developers to transcend the infinite choices created by the competition; indeed, like companies in other industries, developers seek avenues they can use to establish emotional connections with customers, become irreplaceable, and forge enduring relationships. In this regard, a replicable standard no longer signifies a rigid imposition or a destination to be reached by many. Instead, it embodies an assortment of intangible preferences that serve as references for spatial production within a multitude of real estate markets that continually change and evolve.

Architecture: Repetition and Difference

In the architectural discourse, an ongoing tension persists between the imperative for universal forms of modernization and the desire for architectural expression and contextual responsiveness; this forms an enduring and intrinsic dichotomy within the architectural discourse, with deep-rooted origins in the whole of modern architecture. Despite being critiqued from various intellectual standpoints, the concept of architectural standardization gained significant momentum during the modernist movement in the early 20th century. The rise of industrialization, mass production, and prefabrication techniques propelled the integration of standardization principles into architectural practice. It was also politically promoted as a way to provide universally accessible housing. During that period the establishment of certain standards, at least in terms of minimum spaces and basic living infrastructure, ensured a baseline of habitation for all, transforming the architectural standard into both a spatial and political construct. Nonetheless, the advent of neoliberalism, globalization, and market-driven approaches, initially in Western nations and subsequently across the Eastern bloc, reshaped the perception of architectural standardization. It transitioned from an experimental ethos to a more pragmatic orientation in which the replication of successful designs was exploited by real estate developers to achieve economies of scale and expedited construction aimed at reducing costs and market risks. The replication of standard products promised predictability, ostensibly minimizing design errors, and guaranteeing consistent financial outcomes. However, critics have contended that this standardized approach introduces uniformity and monotony into the built environment, stifling creativity and dismantling cultural diversity, thus assigning a mainly negative connotation to architectural standardization.

Nevertheless, as previously mentioned, the construction of new settlements along the BRI is a material witness to the dynamic interplay between multiple and diverse forms of standardization and renewed localized adaptations. The conceptual transition from a rigid notion of standardization to a softer and more immaterial concept can lead to novel interpretations of standardized practices, elements, and forms in architectural design. Although standards still govern spatial production, their influence on the built environment has become more nuanced and diverse. The repetitive application of components, planar arrangements, and even entire structures, no longer results only from adhering to optimized standards; it becomes a practice that reinforces fresh imageries and established ways of life and work.

Seen from this perspective, architectural repetition should not be considered merely as a way to engender homogeneity and monotony. The replication of design elements on different scales and operational levels goes beyond mere duplication. As suggested by French philosopher Gilles Deleuze, repetition serves as a generative and creative energy, giving birth to difference and paving the way for new possibilities.[10] Difference emerges through repetition, ushering in a continuous state of transformation. Embracing standardized practices in architecture thus becomes a creative endeavor, operating across various dimensions, from minute differentiations on the micro level to overarching patterns and structures on the macro level. The generation of differences, particularly through subtle adjustments, fits, and intricate details, becomes the route we can take to break free from hierarchical and binary thinking. It opens up avenues that make it possible to explore a kind of architecture in which standardization and customization are not seen as opposites, but rather as part of the same force that generates the built environment. This is particularly evident in the forms of developments that occur thanks to the circulation of techniques, standards, and imageries, such as those along the Belt and Road Initiative.

While we acknowledge the above, we should also recognize that these developments are neither local nor global, neither contextual nor universal: their nature lies in the multiple understanding of foreign standards and local rules. Indeed, paraphrasing the most innovative research on housing standardization, led by Professor Sam Jacoby at the Royal College of Art in London, we could say that while housing design strategies possess a universal foundation, it is vital we appreciate the fact that these plans also encapsulate socio-technical practices, historical contexts, norms, and expectations.[11] They simultaneously mirror economic constraints and local housing policies, molding the functionality and usability of residences. A thorough analysis of these relations allows us to gain a potent framework for interpreting housing and other architectural facets in transnational development contexts. This approach notably spotlights the dynamic interrelationship between social and spatial practices, one which is often neglected in conventional studies of housing systems.

1 Davide Ponzini, *Transnational Architecture and Urbanism: Rethinking How Cities Plan, Transform, and Learn.* (London: Routledge, 2020). **2** Christopher Alexander and Howard Davis, *The Production of Houses.* (Oxford: Oxford University Press, 1985). **3** Keller Easterling, *Organization Space: Landscapes Highways, and Houses in America.* (Cambridge: The MIT Press, 1999). **4** Fernando Lara, *Global Apartments. Studies in Housing Homogeneity.* (Raleigh: Lulu Press, 2009). **5** Keller Easterling, *Organization Space: Landscapes Highways, and Houses in America.* (Cambridge: The MIT Press, 1999). **6** Keller Easterling, *Extrastatecraft: The Power of Infrastructure Space.* (New York: Verso, 2014). **7** Josh Watkins, "Spatial imaginaries research in geography: Synergies, tensions, and new directions." *Compass* 9, no. 9 (2015): 508-522. **8** Bianca Bosker, *Original Copies: Architectural Mimicry in Contemporary China.* (Honolulu: University of Hawaii Press, 2013). **9** Sharon Zukin, "Reconstructing the Authenticity of Place." *Theory and Society* 40, no. 2 (2011): 161–165. **10** Gilles Deleuze, *Difference and Repetition.* (London: The Athlone Press, 1994). **11** Sam Jacoby, Alvaro Arancibia, and Lucia Alonso. "Space Standards and Housing Design: Typological Experimentation in England and Chile." *The Journal of Architecture* 27, no. 1 (2022): 94–126.

Super Gathering Places. Meeting in between Architecture

The Xi'an Silk Road International Exhibition Center, with its nearly 300-meter-long sides of completely open and free exhibition courts, was designed by the renowned architecture firm GMP on the outskirts of Xi'an, China; the scale of this single building rivals Koolhaas's concept of "bigness."[1] Its complex structural system - two gigantic, symmetrical horizontal elements encircled by a rhythmic repetition of 180 slender steel columns giving the building a sense of lightness - serves as both a means and an end. On the one hand, this framework provides a vast, flexible space free from any obstructing vertical elements in the center, thus transforming the architecture into a container offering diverse possibilities. On the other hand, the structure itself becomes the primary architectural expression, turning structural engineering into a form of artistic expression (see *Xi'an Silk Road International Exhibition Center, Xi'an, China*, p. 194).

Alongside other notable projects such as the Bangladesh-China Friendship Exhibition Center in Dhaka (see *Bangladesh-China Friendship Exhibition Center, Dhaka, Bangladesh*, p. 212), and the Langfang Silk Road International Cultural Exchange Center in China (see *Langfang Silk Road International Cultural Exchange Center, Langfang, China*, p. 206), these monumental structures include an array of spatial arrangements tailored for exhibitions, commercial transactions, cultural events, and showcase settings. Apart from their separate functions, they are envisioned to be places where people can gather. In this respect, they embody the rhetorical essence of the Belt and Road Initiative: an infrastructure of cooperation and exchange. Since the aim of the BRI is to act as a framework for fostering connection, dialogue, and diplomatic endeavors, these buildings serve as pivotal catalysts for the opportunities and interactions that emerge in business and social practices. Consequently, their advanced architecture, structural innovations, and spatial grandeur create the environmental prerequisites required to facilitate such activities.

Their critical urban mass and distinctive structural expressionism trigger a profound question that transcends their functions: should a building exist as a self-contained world, a monument to its own logic and coherence, or should it actively engage with its context, thus shaping and being shaped by its surroundings? This dialectic remains central to the discourse of contemporary architecture, where buildings of such scale often define their own spatial and formal order while simultaneously influencing surrounding environments. In this manner, these superstructures assert their presence as transformative agents within their own urban and cultural landscapes.

Interdisciplinary Practices and Critical Pragmatism

A closer look at the design and construction processes of the large-scale projects, which we identified as "super gathering places," unveil the presence of complexity and interdisciplinarity as a prevailing condition of contemporary architecture. While the investigation of complexity, or as Kiel Moe describes it, "the inherent complexities that characterize contemporary architecture,"[2] is not new, it has gained renewed significance due to the rapid global development of architectural production. Although complexity in architecture can sometimes be simply understood through intricate building forms or in terms of language, its true essence is often challenging to grasp visually, as discussed by Robert Venturi in his seminal work *Complexity and Contradiction in Architecture*.[3] Complexity is more accurately perceived through its defining contingencies, performances, and potential impact.[4]

Adopting this perspective means acknowledging that every architectural project is shaped by a unique combination of theoretical and practical factors that influence both its design and execution. Integrating these varied attributes into the production of built artifacts requires an interdisciplinary approach that introduces a wide range of "hidden agents" into the design process.[5] According to Jane Rendell, such agents seek to question and transform the social conditions of the sites where they intervene, testing and challenging the boundaries of their own disciplines.[6] Large-scale projects within the framework of the BRI often embody this approach since they usually require the establishment of Joint Working Groups made up of experts from different fields such as economics, geography, environmental science, urban planning, security, information technology, and various branches of engineering.

However, when transitioning from an empirical to a theoretical understanding of such architectural productions, the concept of "interdisciplinarity" itself poses certain challenges. On the one hand, scholars such as Jane Rendell and Michael Hays associate interdisciplinarity with the potential for political critique within architectural practice, advocating a form of critical architecture that resists the self-confirming and conciliatory operations of dominant cultural forces.[7] On the other, interdisciplinarity also implies integration between different actors and skills beyond the boundaries of architecture; according to Jeremy Till, these forces often shape architectural outcomes more profoundly than the internal processes of the architect alone.[8]

To unpack this dialectical opposition it is worth considering a distinctive approach that emerged in Chinese architecture in the early 21st century, known as "critical pragmatism." Initially proposed by Li Xiangning during an exhibition at Harvard's Graduate School of Design focusing on the burgeoning practices of young Chinese architects, critical pragmatism advocates for a design philosophy that simultaneously scrutinizes and seeks to transform the social and cultural contexts of architectural outcomes while embracing a pragmatic, problem-solving approach to design that responds to the contingencies of place and time.[9] This intellectual trend strives to integrate critical analysis, practical considerations, and diverse transdisciplinary influences, thus fostering a nuanced architectural approach that effectively addresses the complexities of our contemporary world.

The concept of pragmatism in architecture is not novel. The philosophical approach formulated by American philosophers, such as Charles Sanders Peirce and William

James, emerged in the early 20th century as a reaction against utopian idealism.[10] Reinterpreted in architectural terms, pragmatism emphasizes practicality and the implementation of ideas through action; it advocates an attitude toward design that prioritizes specific contingencies and addresses what works best within a given context. This approach reflects a shift from abstract theoretical concerns to a focus on real-world applications and the effectiveness of architectural solutions in addressing concrete problems. While partially aligned with the post-critical movement in the United States,[11] or Michael Speaks's notion of "design intelligence,"[12] recent interpretations of pragmatism in architecture place greater emphasis on the real-world application of theories, recognizing the inseparable relationship between external forces and human agency behind architectural innovation. In this context, the validity of a critical approach in Chinese architecture relies on its successful testing through practical experimentation and its ability to bring about tangible social progress.

The concept of critical pragmatism was initially adopted to describe the practices of a younger generation of Chinese architects struggling to position their work within the dynamics of globalization and localization, but later gained renewed relevance when it was applied to foreign aid projects. To fully understand the role of criticality and pragmatism in the architecture of the BRI it is crucial to acknowledge that the latter are not entirely new in Chinese architectural practices abroad. The roots of a pragmatic approach, combined with a drive for social change inspired by socialist ideology, can be traced back to China's involvement in global infrastructure development during the Mao era. Scholars such as Charlie Xue,[13] Salvador Santino F. Regilme, and Obert Hodzi have noted that China's engagement in infrastructure and urban development projects on a global scale has long been characterized by a pragmatic approach which is simultaneously symbolic and regionalist.[14] They have differentiated between Western and Chinese architectural aid, highlighting the unique characteristics of Chinese architectural practices, particularly in large-scale construction endeavors. Noteworthy attributes include simplified design, low-cost construction technology, rapid construction processes, and standardized practices.

However, critical pragmatism in the context of Chinese design also involves a refined and continuous negotiation of cultural identities and political strategies. This is achieved by incorporating rich semiotic elements, monumental designs, and bold symmetries that imbue structures with cultural significance and architectural autonomy while addressing practical concerns. This approach is embodied by iconic projects like the Xi'an Silk Road International Exhibition Center or the Lianyungang Industrial Exhibition Center (see *Lianyungang Industrial Exhibition Center, Lianyungang, China*, p. 200); these projects reflect the formal aesthetics of 20th-century Chinese architecture while embracing contemporary materials and technologies that increase the economic and environmental performance of buildings. For example, in the Xi'an Silk Road International Exhibition Center, references to traditional Chinese architecture can be seen in the symmetrical and axial shape of the building, the horizontal proportions of its façades, and the thin columns that surround the structure, filling it with natural light. As in traditional Chinese buildings, these exterior columns are both ornamental and functional, surrounding the central volume along its entire perimeter. As suggested by Andrew W. Charleson, even in architectural projects where structural or functional aspects take precedence, it is crucial to recognize that such architecture extends beyond the mere erection of physical structures.[15] These aspects embody an intricate interplay of cultural symbols, politics, and form, revealing the broader socio-political narratives that shape the built environment.

Indeed, if on the one hand recent Chinese urbanization and its ultimate development in foreign settings has given architects the opportunities and freedom to explore new formal attitudes and technical innovations, on the other hand, as mentioned previously, Chinese politics has inevitably influenced architectural form. This interplay is particularly evident in the architecture of the BRI, where Chinese architects working in collaboration with construction and development firms worldwide embrace new forms of pragmatism as they navigate the challenges of different contexts while aligning with the initiative's broader political objectives. In these projects, the interplay of formal experimentation and the critical inheritance of culture has emerged as a driving force in shaping new architectural expressions.

Tectonic Expressions and Construction Realities

Understanding the architecture of the BRI through the intertwined perspectives of pragmatism and criticism requires a nuanced appreciation of the constructive and technological dimensions underpinning these ambitious projects. This complexity becomes particularly pronounced in large-scale developments, where the processes of construction and on-site implementation are not merely technical concerns, but critical arenas for negotiation, adaptation, and innovation. The management of these processes is deeply integrated with not only the broader financial and institutional framework of the BRI, but also the employment and management of local labor and material supplies in host countries – factors that often become sources of tension and debate. In many instances, local supply chains are deemed insufficient because they do not align with global standards frequently derived from positivist design principles ontologically linked to colonialist ideals.[16] As a result, these projects require complex management frameworks established on-site by leading Chinese design and construction firms; these frameworks are established in order to maintain strict oversight over quality, finances, timelines, and logistical arrangements, while simultaneously adapting to the regulatory requirements and cultural expectations of the host countries.

A case in point is the Bangladesh-China Friendship Exhibition Center designed by the Beijing Institute of Architectural Design in collaboration with China State Construction Engineering Corporation. The project has reflected the emphasis on bilateral cooperation from its inception, being jointly and equally financed by the Chinese Ministry of Commerce and the Bangladeshi Ministry of Finance. This governmental collaboration is mirrored in the sourcing strategies for both materials and labor and aimed at maximizing mutual economic benefits. To achieve this, a robust organizational structure comprising a network of diverse stakeholders was established to navigate the dual challenge of utilizing local resources while importing specialized materials from China (i.e., electrical equipment and high-precision steel elements). An on-site procurement team made up of both Chinese and Bangladeshi project managers was created during the earliest phases of the project to facilitate effective communication between the Chinese design team and local stakeholders, including the project owner, local consultants, government agencies, and regulatory authorities. In turn, this ensured the seamless integration of diverse contributions to the design and construction processes. The centralized procurement team enhanced cost efficiency by implementing bulk orders that were managed via a dedicated online platform; this enabled the realization of a modern and complex structure accommodating a broad array of technological solutions imported from abroad and adopted in a place with scarce resources. The ensuing architecture features a state-of-the-art steel structure; its complex tridimensional trusses support an enormous, suspended, double-curved wavy roof – a strong monumental and visual feature that not only shelters the two-story exhibition hall, but also creates a double-height space filled with natural light, establishing the interplay between interior and exterior (see *Bangladesh-China Friendship Exhibition Center, Bangladesh*, p. 212).

While a similar design and management approach focusing on architectural material qualities and their supply chains has been employed in many BRI projects, the architectural outcomes have varied significantly. For instance, the General Hospital of Niger, designed by CADI Architects in collaboration with the CITIC General Institute of Architectural Design and Research Co. Ltd., faced additional complications due to the lack of advanced construction technologies and compliance with Chinese standards amid local political and economic instability. After a thorough evaluation, the on-site procurement team opted for a pragmatic solution, i.e., utilizing regionally sourced materials, such as local river sand and white cement to reduce costs and logistical burdens. This approach not only addressed technical constraints, but also became a defining architectural feature; it resulted in earthy, ochre-coated façades that blend seamlessly with the arid landscape, thus embodying a tactile connection to the site's natural context.

The architectural expressions of these projects foreground their tectonic characteristics, where structural and material elements take precedence over spatial or functional considerations. In these large-scale buildings, spatial and functional attributes are subordinated to a deeper exploration of how architecture can express its own making. This shift aligns with a tectonic conception of architecture which, as argued by Kenneth Frampton, mediates and enriches the primacy of space through a reconsideration of the constructional and structural methods by which it is realized.[17] In this context, tectonic expression becomes a means with which to express the interplay between materiality, construction, and cultural narratives. Such a tectonic approach imbued in the BRI's projects often appears as a deliberate response to the varied conditions and symbolic demands of these international endeavors.

However, the architecture of the BRI also allows a more multifaceted interpretation of tectonics, one that interweaves the material, social, and syntactic dimensions of architecture. In a recent essay, Anne-Catrin Schultz highlights how the ongoing digitalization of design, fabrication, and construction, along with its expanding global reach, suggests a significant evolution in the understanding of architectural tectonics.[18] This evolution points toward new frameworks that redefine the contemporary role of structural systems, construction processes, and supply chains. In the 21st century the term "digital tectonics" gained prominence thanks to authors such as Jesse Reiser, Nanako Umemoto, and Wassim Jabi; in this context it captures one side of this expanded notion by addressing the intersections between fabrication and the assembly of complex forms, removing the separation between representation and construction.[19]

An example of this evolution is the National Library of El Salvador, the first Chinese foreign aid project to use digital twin technology for its design and construction. The Central-South Architectural Design Institute (CSADI) and the Yanjian Group adopted the French Dassault "3DEXPERIENCE" platform, thus enabling a seamless digital integration of all project phases. This approach facilitated the creation of a digital library of construction components, allowing teams to not only simulate various design options, but also control supply chain logistics, as well as optimize costs from conception to realization. By employing a digital twin, the design team could adapt the project in real time, thus ensuring that the materials sourced – whether local or imported – were both cost-effective and of high quality. The platform also enabled the customization of complex components and their on-site three-dimensional delivery using a laser pointer; this allowed for the construction of the library's distinctive double-curved façade without the need for specialized labor, thus leveraging the local workforce (see *National Library of El Salvador, San Salvador, El Salvador*, p. 218).

In this instance, the construction and design processes were anchored by a unified data model, enabling high-precision manufacturing and seamless assembly of architectural components. Rather than adhering to Frampton's notion of a "textile interpretation of tectonics," these architectural models reflect a digital reconfiguration of material elements, from early project conception to on-site realization. The fact that many of these components, akin to those used in manufacturing industries, are prefabricated, is a secondary issue. Instead, what is truly significant is the development of digital artifacts that effectively bridge the gap between global design practices and local construction realities.[20]

These examples reflect a broader transformation in architectural practice within the BRI framework: a move away from purely spatial and functional definitions of architecture and toward a multifaceted understanding of tectonics that embraces both the immaterial and material realities of construction. The shift to digital fabrication and global procurement networks introduces new paradigms for architectural production, one where the distinction between design and construction becomes blurred, and the role of the architect extends into realms traditionally dominated by engineering and logistics. It emphasizes a transversality that moves beyond conventional disciplinary boundaries. Architecture becomes a logistic stance, one where, in line with the principles of new materialism, human agency is diminished and the focus shifts to how various networks of material and non-material agents converge to collectively shape the built environment.[21] As pointed out by Winifred Elysse Newman, a renewed understanding of architecture through the lens of tectonics can provide a new junction between the imaginary and the real in architecture, challenging those categorical distinctions between rational and sensual, form and matter, and material and virtual.[22]

Architecture: Autonomy and Dependency

As previously discussed, these "super gathering places" represent a mix of advanced construction techniques, driven primarily by China's technological expertise and symbolic ambitions, and practical design solutions tailored to the specific needs and cultural contexts of each project. By integrating cutting-edge digital tools and management practices with both local and global supply chains, these projects exemplify how architectural practices can operate at the intersection of multiple disciplines and design approaches.

The architectural approach adopted in these projects is dual in nature; it is shaped by both the aesthetic aspirations of global design and the pragmatic demands of construction in different local settings. This duality challenges two dominant theoretical positions in architecture which, as suggested by K. Michael Hays, have historically defined the field.[23] On the one hand, architectural artifacts are seen as instruments of a self-sustaining cultural hegemony that perpetuates its own values. On the other, architectural objects are understood as isolated containers of internal coherence, created through the agency of the architect, detached from external forces, and focused purely on formal considerations.

Our nuanced understanding of architectural tectonics and criticism aims to dismantle the dialectics that often underlie Western approaches to architectural categorization. Instead, it advocates an approach that embraces the inherent complexity of architectural production, recognizing the systemic nature of integrated practices that define contemporary global construction endeavors. Following the perspective of historian Carlo Olmo, we argue for an architectural language inspired by the inherent contradiction between intellectual labor and the material forces of production.[24] It is within this tension, i.e., between the designer's intentions and the contextual contingencies of a given project, that true architectural innovation can emerge.

In other words, we advocate a renewed focus on theoretical and design perspectives that bridge the gap between autonomy and dependency. This kind of approach, that transcends binary thinking, would not only yield a more comprehensive understanding of architectural artifacts, but would also enable a deeper analysis of buildings as they engage with diverse cultural, social, and economic realities.[25] This perspective positions architecture as a critical practice that must actively negotiate the complexities of its context rather than merely exist as a purely self-contained discipline.

However, this approach also means moving beyond the conventional idea of interdisciplinarity, which often involves the superficial borrowing of concepts from other fields without fully integrating them into architectural practice. In contrast, Marianna Charitonidou proposes an alternative epistemological model based on Deleuze and Guattari's concept of *transversality*.[26] She suggests that, unlike interdisciplinarity, transversality rejects the false dichotomy between content and context, as well as the division between architecture and society. It acknowledges the dynamic tensions between autonomy and heteronomy, between generalization and specificity, and between art and science.[27] These tensions, rather than being resolved dialectically, can instead fuel a more nuanced and adaptable architectural practice. In this framework, architectural practice becomes non-categorical and non-judgmental. It defies disciplinary categories and resists hierarchies, encouraging architects to engage with the full complexity of their projects through integrated practices and design actions.

---

1 Rem Koolhaas, Bruce Mau, Jennifer Sigler, and Hans Werlemann, *Small, Medium, Large, Extra-Large (S,M,L,XL)*. (New York: The Monacelli Press, 1995). **2** Kiel Moe, *Integrated Design in Contemporary Architecture*. (New York: Princeton Architectural Press, 2008). **3** Robert Venturi, *Complexity and Contradiction in Architecture*. (New York: The Museum of Modern Art, 1977). **4** Kiel Moe, *Integrated Design in Contemporary Architecture*. (New York: Princeton Architectural Press, 2008). **5** Tom Avermaete, "Death of the Author, Center, and Meta-Theory: Emerging Planning Histories and Expanding Methods of the Early 21st Century," In Carola Hein, ed. *The Routledge Handbook of Planning History*. (New York: Routledge, 2017). **6** Jane Rendell, "Foreword," in Jonathan Bean, Susannah Dickinson, and Aletheia Ida, eds. *Critical Practices in Architecture: The Unexamined*. (Newcastle upon Tyne: Cambridge Scholars Publishing, 2020): xi-xix. **7** K. Michael Hays, "Critical Architecture – Between Culture and Form," *Perspecta* 21 (1984): 14-29; Jane Rendell, "A Place Between Art, Architecture and Critical Theory," *Proceedings to Place and Location* (Tallinn, 2003): 221-233. **8** Jeremy Till. *Architecture Depends*. (Cambridge: The MIT Press, 2009). **9** See Li Xiangning, *Contemporary Architecture in China. Towards a Critical Pragmatism*. (Mulgrava: Images Publishing, 2018). **10** 'Pragmatism', in *The Stanford Encyclopedia of Philosophy*, ed. Edward N. Zalta (Stanford: Stanford University Press, 2021). **11** See for instance Robert Somol and Sarah Whiting, "Notes Around the Doppler Effect and Other Moods of Modernism," *Perspecta* 33 (2002): 72–77. **12** Michael Speaks, "Design Intelligence: Part 1, Introduction," *A+U* (2002): 10–18. **13** See for instance Charlie Xue's essay in this book. **14** Salvador Santino F. Regilme and Obert Hodzi, "Comparing US and Chinese Foreign Aid in the Era of Rising Powers." *The International Spectator* 56, no. 2 (2021): 114–131. **15** Andrew W. Charleson, *Structure as Architecture*. (Burlington: Elsevier, 2005). **16** Francesco Carota and Giulia Montanaro, "Navigating the Nexus of Global Design Practices and Local Construction Realities. An Exploration into the Collaborative Architectural Endeavors within the Framework of the Belt and Road Initiative (BRI)." In Jiawen Han, Davide Lombardi, and Alessandro Cece, eds. *Advances in the Integration of Technology and the Built Environment: Select Proceeding of AAB 2024*. (Singapore: Springer Nature, in press). **17** Kenneth Frampton, *Studies in Tectonic Culture: The Poetics of Construction in Nineteenth- and Twentieth-Century Architecture*. (Cambridge: The MIT Press, 1996). **18** Anne-Catrin Schultz, "Architectural Tectonics in the Age of Climate Crisis, Social Change and Digital Fabrication," *TAD* 7, no. 1 (2023): 2-3. **19** Jesse Reiser and Nanako Umemoto, *Atlas of Novel Tectonics*. (New York: Princeton Architectural Press, 2006); Wassim Jabi, "Digital Tectonics: The intersection of the physical and the virtual," *Proceedings of the 24th Annual Conference of the Association for Computer Aided Design in Architecture (ACADIA) and the 2004 Conference of the AIA Technology in Architectural Practice Knowledge Community* (2004): 256-269. **20** Francesco Carota and Giulia Montanaro, "Navigating the Nexus of Global Design Practices and Local Construction Realities. An Exploration into the Collaborative Architectural Endeavors within the Framework of the Belt and Road Initiative (BRI)." In Jiawen Han, Davide Lombardi, and Alessandro Cece, eds. *Advances in the Integration of Technology and the Built Environment: Select Proceeding of AAB 2024*. (Singapore: Springer Nature, in press). **21** Ibid. **22** Winifred Elysse Newman, "Tectonics in Equipoise," *TAD* 7, no. 1 (2023): 1. **23** K. Michael Hays, "Critical Architecture – Between Culture and Form," *Perspecta* 21 (1984): 14-29. **24** Carlo Olmo, "One History, Many Stories." *Casabella* 59, no. 619-620 (1995): 75-86. **25** K. Michael Hays, "Critical Architecture – Between Culture and Form," *Perspecta* 21 (1984): 14-29. **26** Marianna Charitonidou, "Revisiting the debate around autonomy in architecture. A genealogy." In Federico Soriano, ed. *critic|all II International Conference on Architectural Design & Criticism. Actas Digitales, Digital Proceedings*. (critic|all PRESS, 2016): 99–127. **27** Ibid.

A Visual Tour of the BRI:
Photographic Reportages

Pakistan-China Technical and Vocational Institute

Photographic Reportage by Al Yousuf

Main entrance to the Pakistan-China Technical and Vocational Institute characterized by Arabic ornamentations, Gwadar, Pakistan.

援巴基斯坦瓜达尔职业技术学校项目
Jawan
Hunarmand Pakistan

Main courtyard surrounded by arched porches in the Pakistan-China Technical and Vocational Institute, Gwadar, Pakistan.

Outdoor hallway of the Pakistan-China Technical and Vocational Institute, Gwadar, Pakistan.

Laboratory room characterized by lattice windows with Arabic ornamentation in the Pakistan-China Technical and Vocational Institute, Gwadar, Pakistan.

A regular grid provides rhythm to the façade of the Pakistan-China Technical and Vocational Institute, Gwadar, Pakistan.

Lianglu-Cuntan Free-Trade Port Area

Photographic Reportage by Raul Ariano

Concrete piers on the Yangtze river charachterize the Lianglu-Cuntan Free-Trade Port Area in Chongqing, China.

Main vehicular entrance to the Lianglu-Cuntan Free-Trade Port Area, Chongqing, China.

The Lianglu-Cuntan Free-Trade Port Area within the urban landscape, Chongqing, China.

Cargos move in front of the Lianglu-Cuntan Free-Trade Port Area, Chongqing, China.

Main pedestrian entrance of the Lianglu-Cuntan Free-Trade Port Area, Chongqing, China.

Kilamba Kiaxi Housing Estate

Photographic Reportage by Ivo Tavares Studio

Repetition of mid-rise and high-rise buildings in Kilamba Kiaxi Housing Estate, Kilamba, Angola.

Natural landscape surrounding the new development of Kilamba Kiaxi Housing Estate, Kilamba, Angola.

Communal space in front of new residential buildings within Kilamba Kiaxi Housing Estate, Kilamba, Angola.

New streets and communal gardens in Kilamba Kiaxi Housing Estate, Kilamba, Angola.

Repetition of new residential buildings in Kilamba Kiaxi Housing Estate, Kilamba, Angola.

Relationship between the building's ground floor and the communal garden in Kilamba Kiaxi Housing Estate, Kilamba, Angola.

Relationship between the residential buildings and the surrounding natural landscape of Kilamba Kiaxi Housing Estate, Kilamba, Angola.

Private balconies of high-rise buildings in Kilamba Kiaxi Housing Estate, Kilamba, Angola.

Lattice wall covering the distribution core of high-rise buildings in Kilamba Kiaxi Housing Estate, Kilamba, Angola.

Ground floor of high-rise buildings in Kilamba Kiaxi Housing Estate, Kilamba, Angola.

Xi'an Silk Road International Exhibition Center

Photographic Reportage by CreatAr Images

The Xi'an Silk Road International Exhibition Center within the urban landscape, Xi'an, China.

Grandeur of the main hall of the Xi'an Silk Road International Exhibition Center, Xi'an, China.

External structure made of two symmetrical horizontal vaults and several columns of the Xi'an Silk Road International Exhibition Center, Xi'an, China.

The Xi'an Silk Road International Exhibition Center within the surrounding urban landscape, Xi'an, China.

The Xi'an Silk Road International Exhibition Center within the surrounding urban and natural landscape, Xi'an, China.

Main entrance under the suspended structure of the Xi'an Silk Road International Exhibition Center, Xi'an, China.

Suspended internal connection between the two cores of the Xi'an Silk Road International Exhibition Center, Xi'an China.

Mullion curtain wall of the Xi'an Silk Road International Exhibition Center, Xi'an, China.

Big square in front of the main entrance to the Xi'an Silk Road International Exhibition Center, Xi'an, China.

II.
Architectural Guide to the Belt and Road Initiative

Gift Complexes

Hybridizing Extra-State Architecture

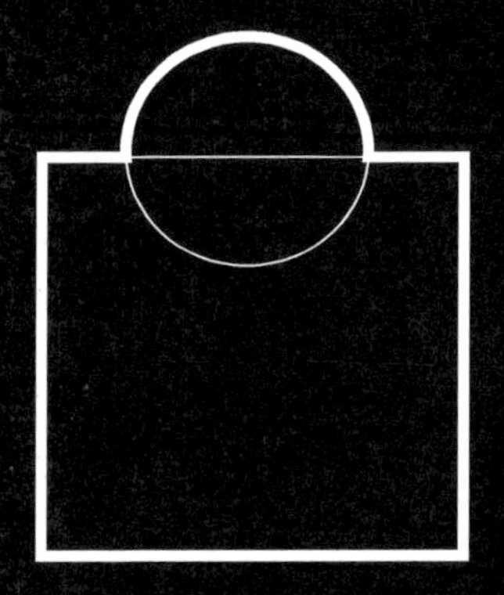

Lanzhou New Area Amusement Park

Location	Lanzhou New Area, Lanzhou, China
Area	313 hectares
Date	2016 - ongoing
Client	Lanzhou Investment Group Co. Ltd.

Hard infrastructures are superimposed on eccentric leisure amenities composing the amusement park.

The Lanzhou New Area, geographically located at a strategic intersection of the Lanzhou-Yinchuan-Xining delta region, stands on a sandy plain in an area once occupied by mountains. Prior to the construction of the new city, 6,000 workers and more than 3,000 excavators were employed to flatten 700 mountains in order to make the over 821 square kilometer area suitable to accommodate a massive urban development. As the inaugural state-level new development zone in the northwestern region of China, Lanzhou is one of the country's strategic initiatives to extend the reach of China's key industries (e.g., high-precision technologies, heavy manufacturing, and chemical production) toward the western regions. The zone includes business parks, residential developments, and leisure amenities together with a new high-speed railway station and an international airport. The Lanzhou New Area Amusement Park, inaugurated in October 2018, stands out as an eccentric location in this area. Indeed, even though a theme park is an almost necessary facility for every new urban development in China, in this case the scale of the project, covering an astonishing 313 hectares, makes it a unicorn throughout the country, not only as a local, but also national attraction.

The area is divided into two main sections. The right side of the plot hosts the "Dinosaurs Park Resort," an amusement park for families including hotels, resorts, water slides, and a commercial street. The biggest landmark in this area is an enormous, 50-meter-high geodesic dome with a diameter of 200 meters. The dome – a self-supporting glass shell structure made up of interlocking steel triangles – functions as an enormous greenhouse in the middle of the desert. Indeed, inside the dome the climate is strictly controlled to provide a faithful reproduction of a tropical environment. The internal space is designed to resemble the landscape of Hainan Island, a famous tourist destination in southern China, and is dotted with tropical plants and water attractions.

The left side of the park, next to the high-speed railway station, is home to the "Great Wall Western Film and Television City," a park where replicas of Chinese and Western monuments rise up from the desert ground. Reproductions of the Temple of Heaven and the Forbidden City in the center of Beijing are visible when approaching the site from the north, right in front of the new high-speed railway station. It also includes a sphinx, the Parthenon, and a turquoise-tiled pavilion covered in arabesque patterns, reminiscent of the Jameh Mosque in Isfahan. The very essence of "replication," named by Bianca Bosker "duplitecture" is evident here. The "culture of the copy," rooted in traditional Chinese philosophy, is embodied in the architectural approach; all the artifacts in the park are reproduced on a 1:1 scale with minute attention to detail.

The whole area seems designed to represent the quintessential rhetoric of the Belt and Road Initiative, a sphere of development and collaboration connecting different countries around the world.

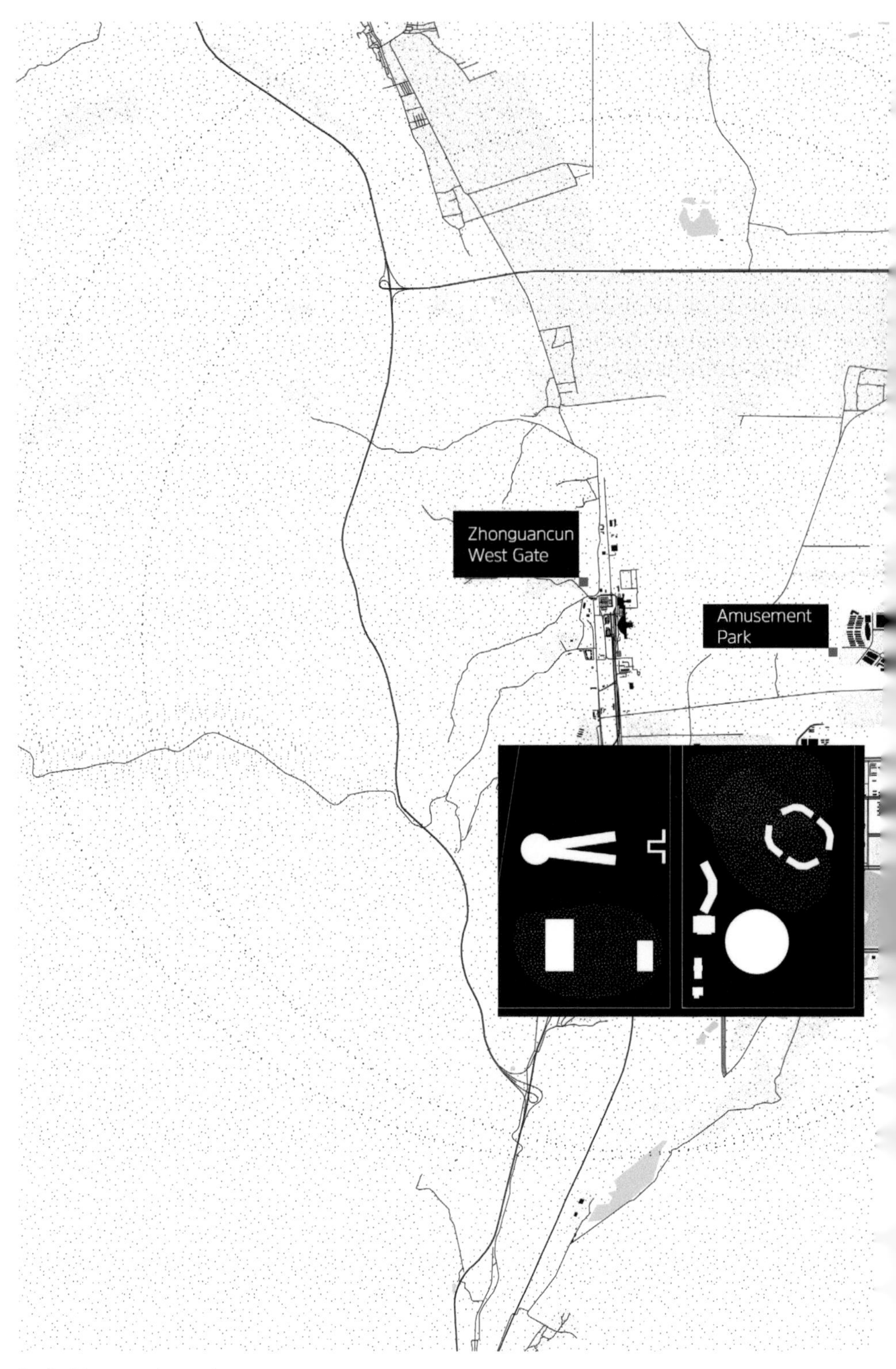

Territorial map on two scales.

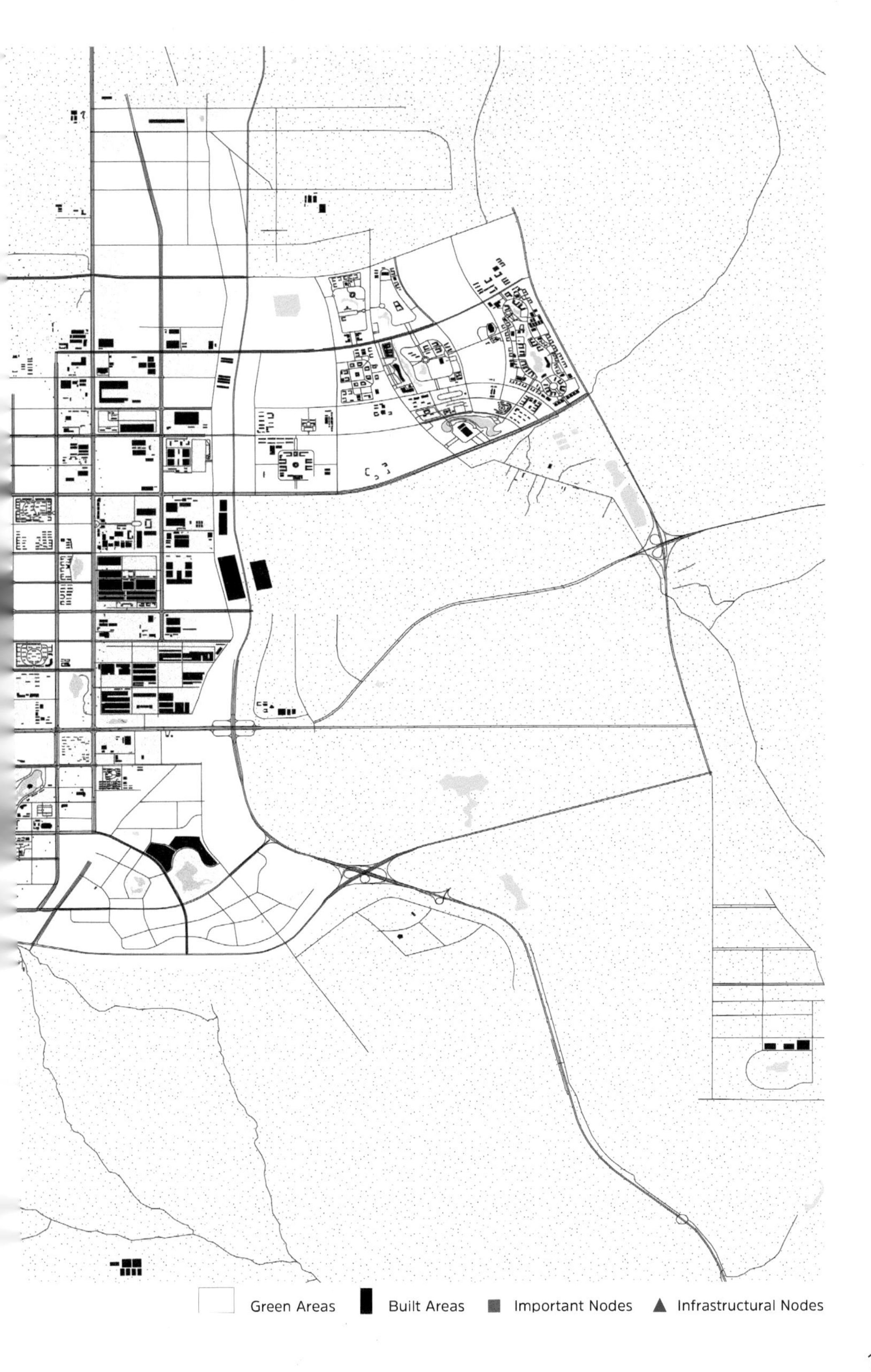
Green Areas
Built Areas
Important Nodes
Infrastructural Nodes

Geodesic Sphere

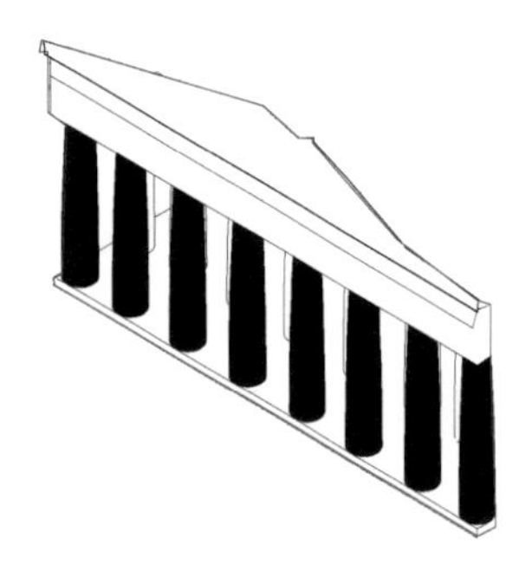

Parthenon

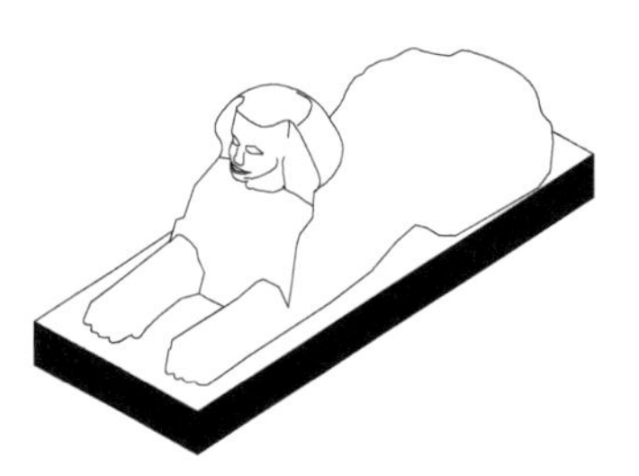

Egyptian Sphinxes

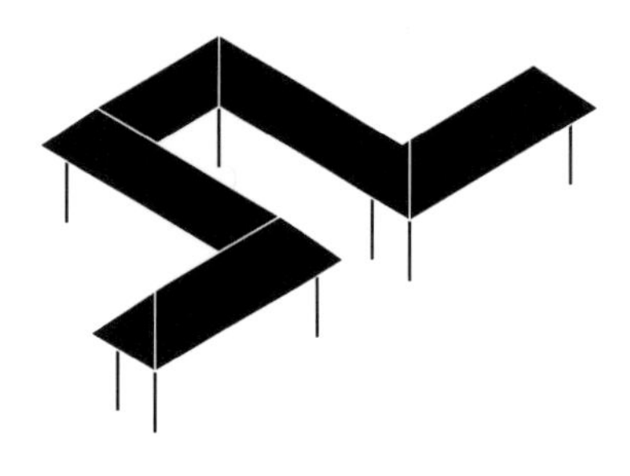

Porticos

Morphological diagram showing the main elements composing the architectural language.

Axonometric view.

Sino-Italian Cultural Exchange City Reception Center

Location	Chengdu, China
Area	2,050 square meters
Date	2021
Architect	Architecture Office Firm (AOE)
Client	Tianfu Investment Group Co., Ltd

Circular walls, arcades, and skylights reminiscent of Classical aesthetics compose the Italian pavilion.

The Sino-Italian Cultural Exchange City Reception Center is a captivating architectural complex located in the dynamic Tianfu New District, next to the bustling city of Chengdu. The district, approved in October 2014 with an envisaged surface area of 1,578 square kilometers, was defined by General Secretary Xi Jinping as "one of the most important nodes in the construction of the Belt and Road." As a result, the Sino-Italian Cultural Exchange City Reception Center, completed in 2021 by the Beijing-based design firm AOE, rapidly became a city landmark; it not only rhetorically represents the union between East and West thanks to its eclectic architectural language, but also reflects the strong cultural exchange that exists between China and Italy. The reception center is a reminder of the historical interactions that already existed between the two countries due to its location: Chengdu was visited by the explorer Marco Polo during his 13th-century journey along the ancient Silk Road.

The project, covering an area of 2,050 square meters, is made up of two distinct yet interconnected sections: the Italian pavilion that acts as an art hall built on the west side of the master plan, and the Chinese cultural hall on the east side. Anyone crossing the center will encounter the striking water feature inbetween these two pavilions which are, however, connected by a long curved corridor linking the two pavilions. The use of this architectural element symbolizes the metaphorical bridge between the two cultures. It is not a coincidence that the famous Anshun Langqiao Bridge was one of the main elements noted and described by Marco Polo during his sojourn in Chengdu. The two parts of the center feature typical elements of both Italian classical architecture and Chinese traditional architecture; together they generate an architectural pastiche in which historical elements are juxtaposed against contemporary reinterpretations. Three main elements dominate the Italian pavilion area: the fountain, the plaza, and the theater. The overall design pays homage to classical Italian architecture; it features circular and semicircular walls and windows that vary in size, echoing the Renaissance theme of circular perfection. The exhibition spaces are bathed in natural light streaming through skylights, creating an enchanting ambiance reminiscent of classical aesthetics. When walking up the three floors that follow the natural lie of the land, visitors will be delighted to discover a roof garden, accessible via a gracefully curved staircase, conjuring up the spirit of Renaissance typologies. On the other side, the Chinese pavilion, which serves as a cultural hall, is a smaller hidden gem set amid a peaceful bamboo forest; it reflects the profound relationship between humans and nature, as emphasized in Feng Shui philosophy. The architectural design is also inspired by traditional Chinese courtyard houses featuring elegant gable roofs, inner courtyards, and circular openings reminiscent of the classical gardens of Suzhou. Here, the theme of the circle is subtly expressed through voids that lead to different spaces and an intricately detailed outdoor pergola.

The center is characterized by a remarkable merging of architectural elements and a seamless intertwining of architectural forms and languages, epitomizing the enduring collaboration between China and Italy.

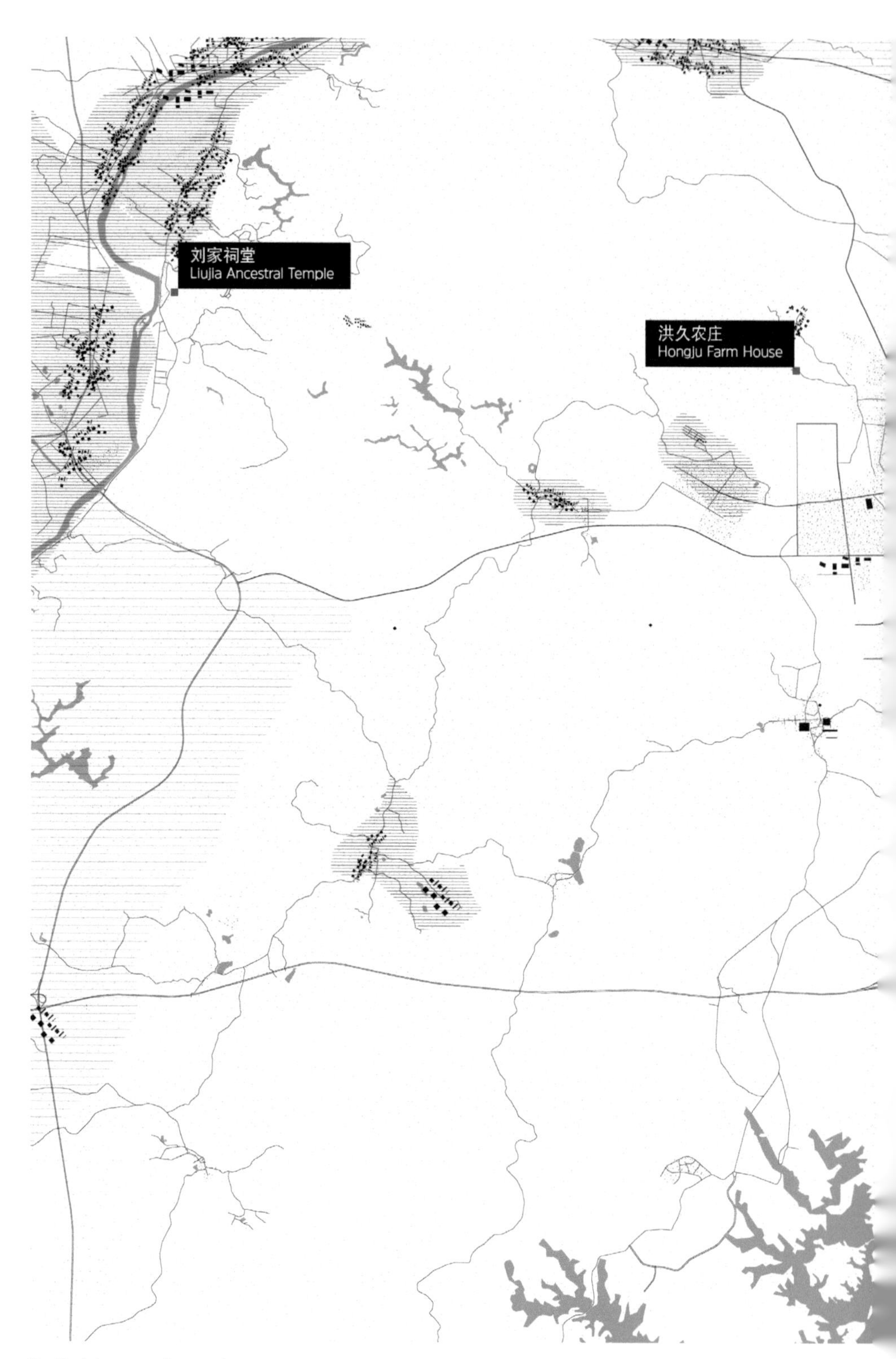

Territorial map on two scales.

General Hospital of Niger, Niamey, Niger
Lanzhou New Area Amusement Park, Lanzhou, China
0
1
2 km
Green Areas
Built Areas
Important Nodes
Infrastructural Nodes

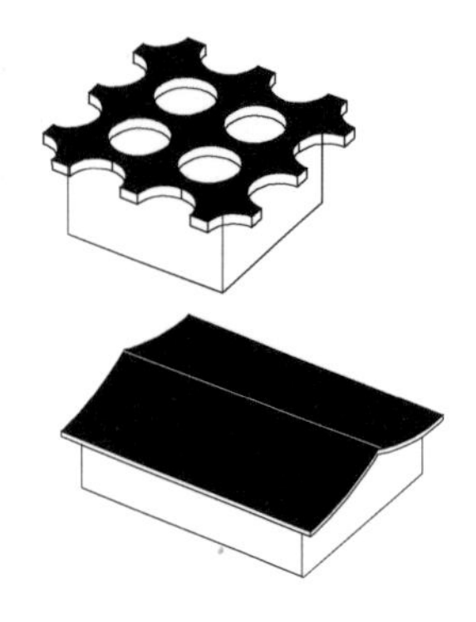

Chinese-style gable roofs

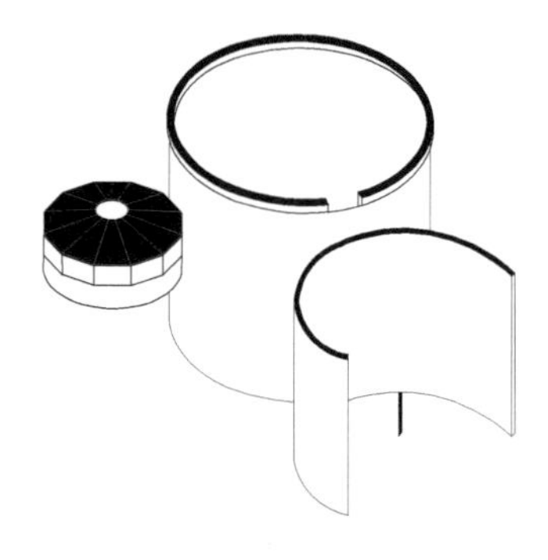

Italian Renaissance spiral staircases
and circular skylights

Chinese Huabiao-style column

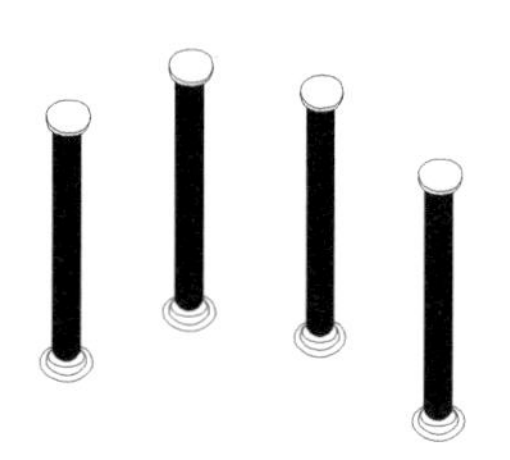

Italian Renaissancestyle rounded columns

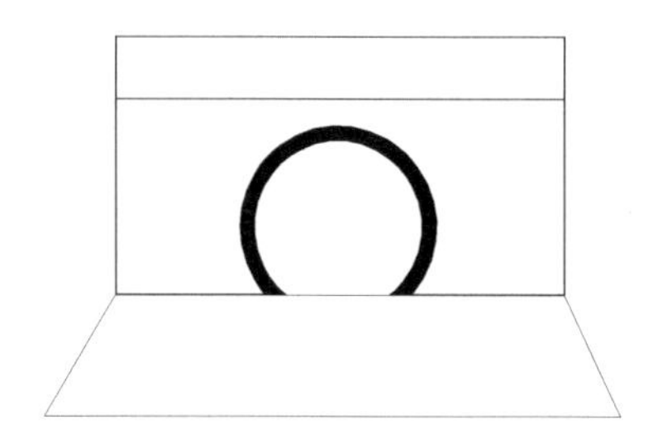

Chinese circular openings in Suzhou style

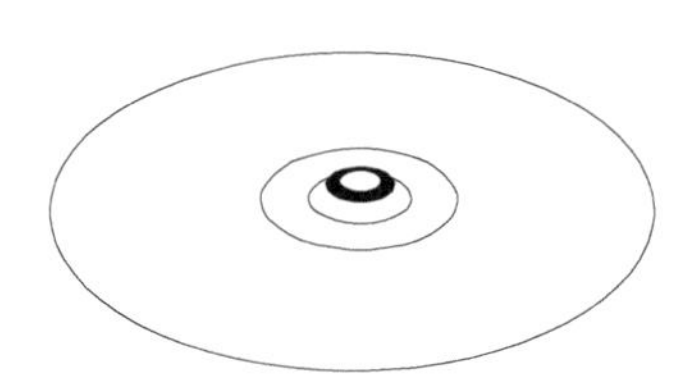

Italian fountain in Renaissance style

Morphological diagram showing the main elements composing the architectural language.

Axonometric view.

General Hospital of Niger

Location	Niamey, Niger
Floor Area	34,000 square meters
Date	2016
Architect	CADI
Client	The Ministry of Commerce of China, Ministry of Public Health of the Republic of Niger

The main entrance is characterized by square-based columns connected to a perforated roof by trapezoidal elements.

The General Hospital of Niger, located on a flat sandy terrain 13 kilometers north of the capital city of Niamey, is one of the biggest and most advanced hospitals in West Africa. Since the turn of the century China's investments in Niger have boosted its development and acted as a form of economic gain for Chinese companies active abroad. The Niamey General Hospital is certainly a successful example and a case in point. The complex was designed by a partnership between the young firm CADI Architects and the big integrated design CITIC General Institute of Architectural Design and Research Co. Ltd. The long, complex phase involving the design of the hospital began in 2010; the building was finally completed in 2016.

With a total gross floor area of over 34,000 square meters, the hospital is made up of fifteen different pavilions. Specific functions are allocated to the individual buildings that are between two and five stories high and are organized around a series of courtyards connected by an intricate but rational system of open-ended walkways and galleries. In distributive and volumetric terms, the General Hospital of Niger embodies all the standard characteristics of a traditional 'linked pavilion' hospital typology. The architectural language is characterized by bold light and shadow effects and earthy ochre-colored façades that make the building merge seamlessly with the surrounding arid landscape. The construction of the exterior walls is based on a local traditional process called the "Tyrol" style; it includes the use of local materials such as river sand and white cement in order to lower the economic and environmental impact of imported materials.

Adapting the General Hospital in Niger to the local context took place not only as regards its architectural language and materials, but also its program and environmental strategies. Firstly, since Islam is the dominant religion in Niger, Muslim worship halls are present throughout the hospital pavilions, thus enriching the general design with unique spaces that not only function as worship halls, but can be turned into temporary camping sites for patients and their families. Secondly, several passive internal comfort control strategies were adopted to deal with the arid and hot climate of Niamey that reaches 43 degrees Celsius during the warmest days of summer. For instance, the distributive internal hallways, located at the perimeters of the pavilions, are open-ended on both sides to allow for natural cross ventilation. Several different sunshade devices have been integrated into the façade in order to mitigate direct solar exposure, thus enabling the microcirculation of air and dissipation of heat while creating an evocative interplay of light and shadow in the corridors. Finally, the prefabricated concrete block roof is equipped with layers of thermal insulation to reduce heat transmission.

The hospital is a successful example of a public health infrastructure implemented thanks to the mutual effort of the governments of China and Niger; its building type is not part of the institutional typologies, such as conference centers, but instead embraces the needs of local communities by adapting to the local climate and geographical context.

Territorial map on two scales.

Pak-China Friendship Center, Islamabad, Pakistan
Koira Tagui
Gendarmerie
Dan Zaima Koira
Green Areas
Built Areas
Important Nodes
Infrastructural Nodes

Courtyards surrounded by porches in Chinese style

"Tyrol"-style openings

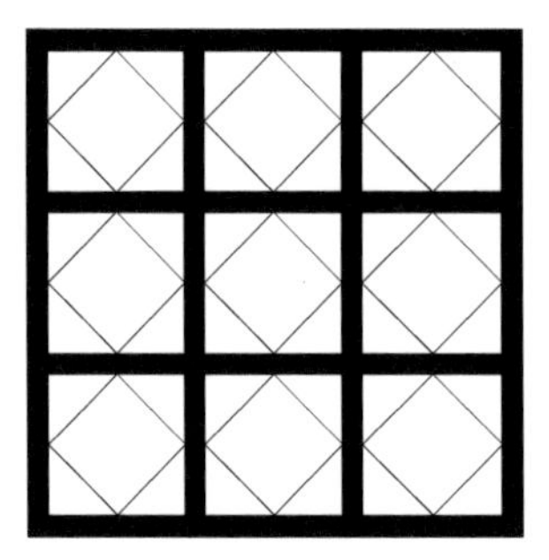

Traditional Niger lattice ornamentation

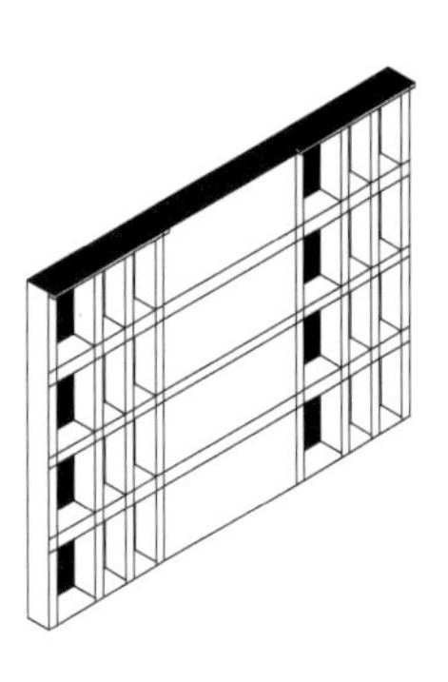

"Tyrol"-style shading devices

Morphological diagram showing the main elements composing the architectural language.

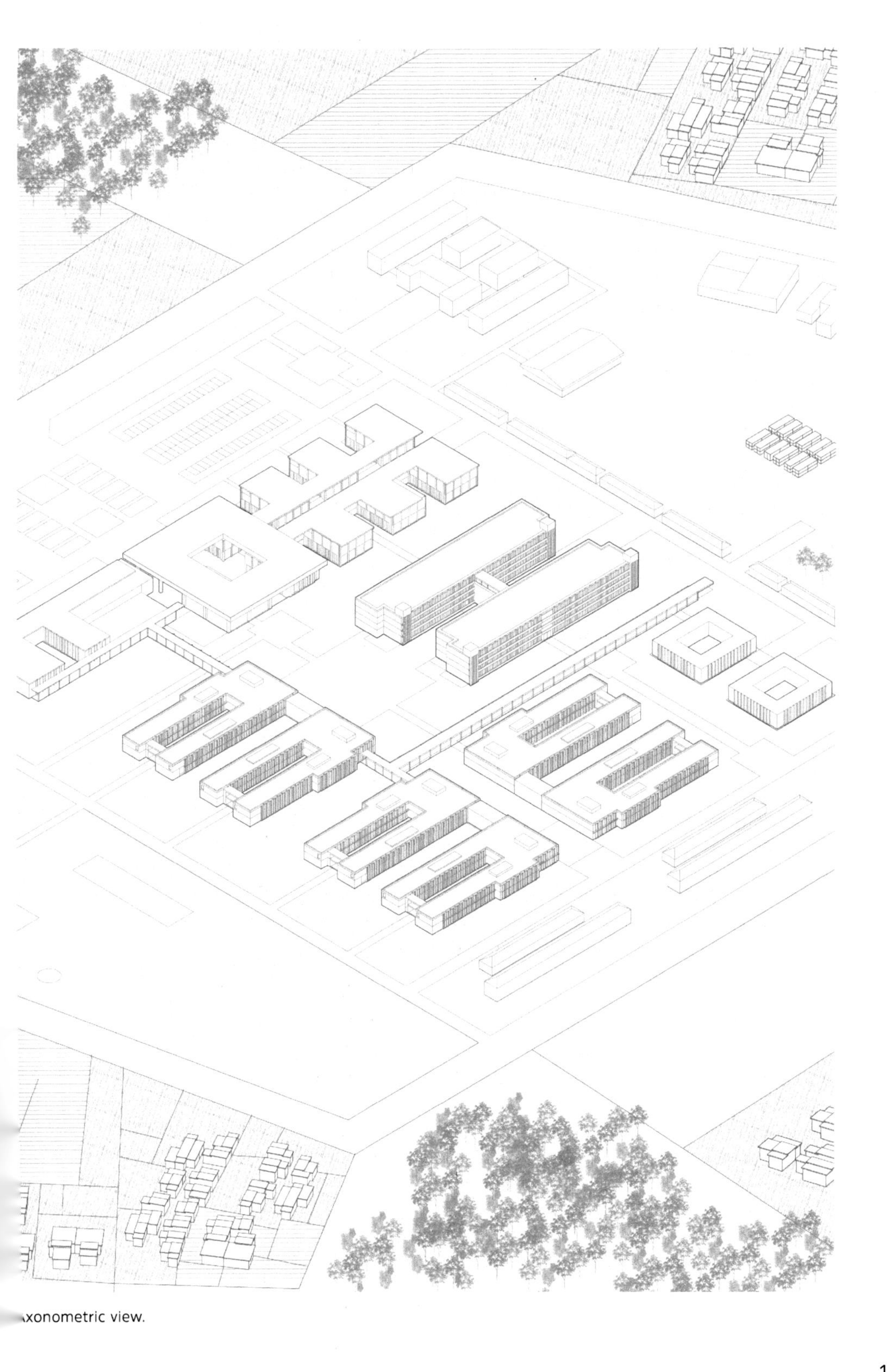

Axonometric view.

Pakistan-China Technical and Vocational Institute

Location	Gwadar, Pakistan
Floor Area	7,350 square meters
Date	2021
Architect	China Communications Construction Company (CCCC)
Client	Pakistan's Ministry of Planning Commission of China, Ministry of Public Health of the Republic of Niger

The main façade is characterized by a dome and a repetition of rounded arches inspired by the Gwadar vernacular mud architecture.

In October 2021 the Pakistan-China Technical and Vocational Institute (part of the China–Pakistan Economic Corridor, CPEC) was inaugurated in the city of Gwadar, Balochistan Province. Over the years China has been a crucial partner for Pakistan, providing investments in numerous areas such as energy, telecommunications, and comprehensive infrastructure developments. With a whopping 54 billion US dollars spent by China on infrastructure, Pakistan emerged as the largest recipient of investments within the Belt and Road Initiative. The projects completed under the CPEC agreement also include the Development of the Port and Free Zone, Gwadar Smart, the Port City Master Plan, and the Gwadar East Bay Expressway. The Pakistan-China Technical and Vocational Institute is a technical training institute that provides much-needed training and education facilities. One of its functions is to offer specialized skills to locals so they can qualify for jobs at Gwadar Port, under expansion as part of the China–Pakistan Economic Corridor.

The total built area of the institute is around 7,350 square meters; the two-story buildings accommodate specialized workshops, laboratories with advanced simulation instruments, multi-functional halls, as well as classrooms and dormitories for students and faculty members. The institute's design combines two main wings connected by a welcoming porch. The U-shaped layout of the center embraces both functionality and the search for monumentality. The building is organized as a series of subsequent courtyards and spaces surrounded by porches – a clear reference to the Chinese vernacular architecture of traditional houses, namely "Siheyuan"; it is an ancient, but efficient spatial strategy for integrating the building with the natural environment and blurring the boundaries between indoor and outdoor spaces and activities. The building's monumentality is accentuated by the main façade where the use of a central pointed arc and a dome drew on the local architectural language. The façade design is based on a bold vision that embraces Middle Eastern architectural elements, incorporating the intricate style of Arabic lattice ornamentations and the grace of rounded arches, inspired by Gwadar vernacular mud architecture. The lattice wall was carefully designed to create the interplay of sunlight and shade characterizing the atmosphere of the hall of the main building.

Given the lack of a proper modern construction industry in the city of Gwadar, CCCC opted for a simple construction system made of prefabricated concrete elements which, together with all the other buildings components such as electrical equipment, appliances etc., are imported from China and assembled in Gwadar. While this contingent factor is only partially visible in the exterior of the building characterized by simple forms and an austere appearance, it is much more evident in its interior. Indeed, apart from the grand entrance lobby, all the institute's interior spaces are furnished and decorated with elements imported from China, thus giving them an unmistakably exotic taste.

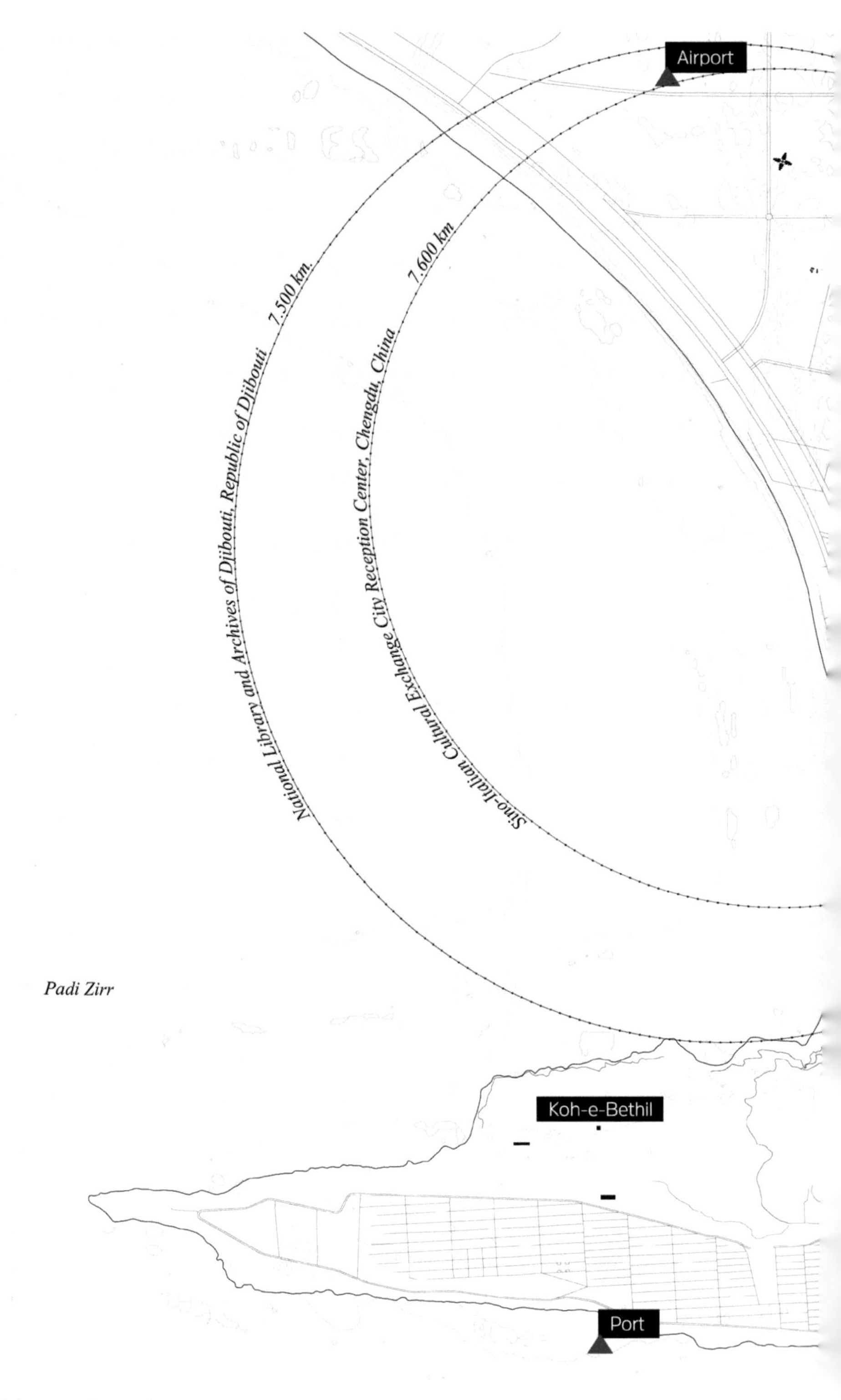

Territorial map on two scales.

50m
25
Demi Zirr
Akram
10 km
5
0
Green Areas
Built Areas
Important Nodes
Infrastructural Nodes

Rounded arches inspired by the Gwadar vernacular mud architecture

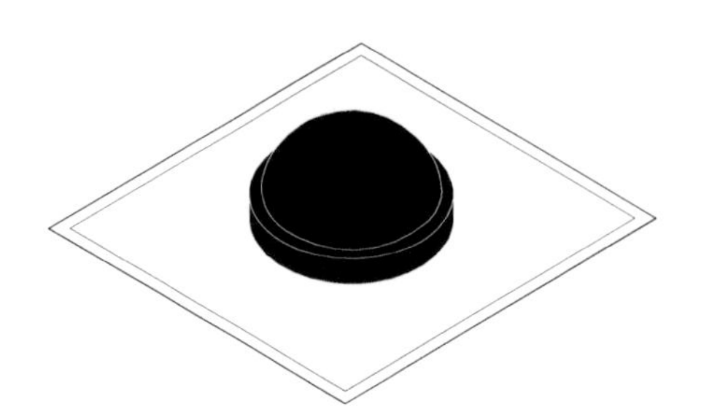

Islamic-style spherical dome

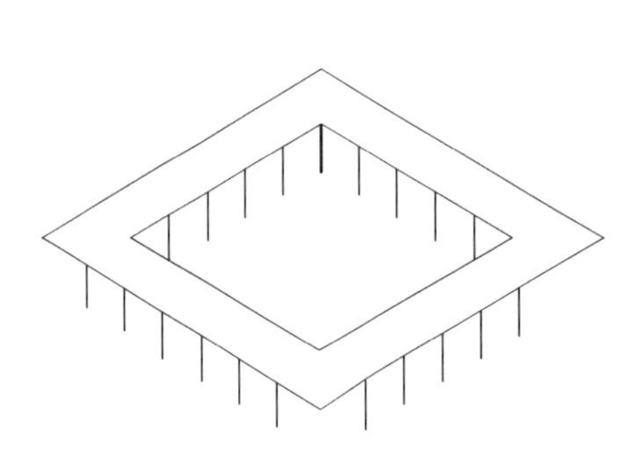

Courtyards surrounded by porches in Chinese style

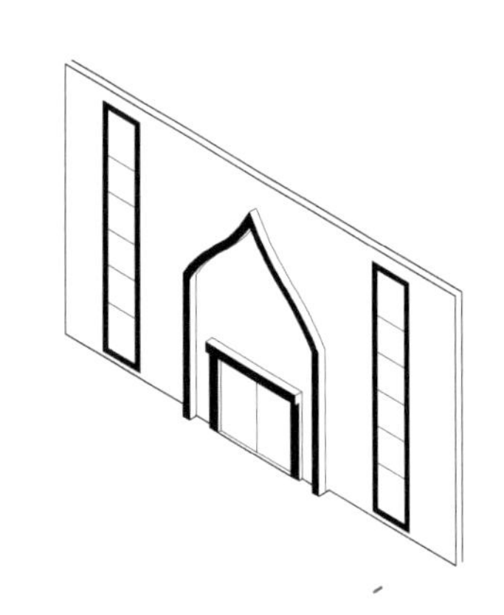

Pointed arches and ornamental lattice in Islamic-style

Morphological diagram showing the main elements composing the architectural language.

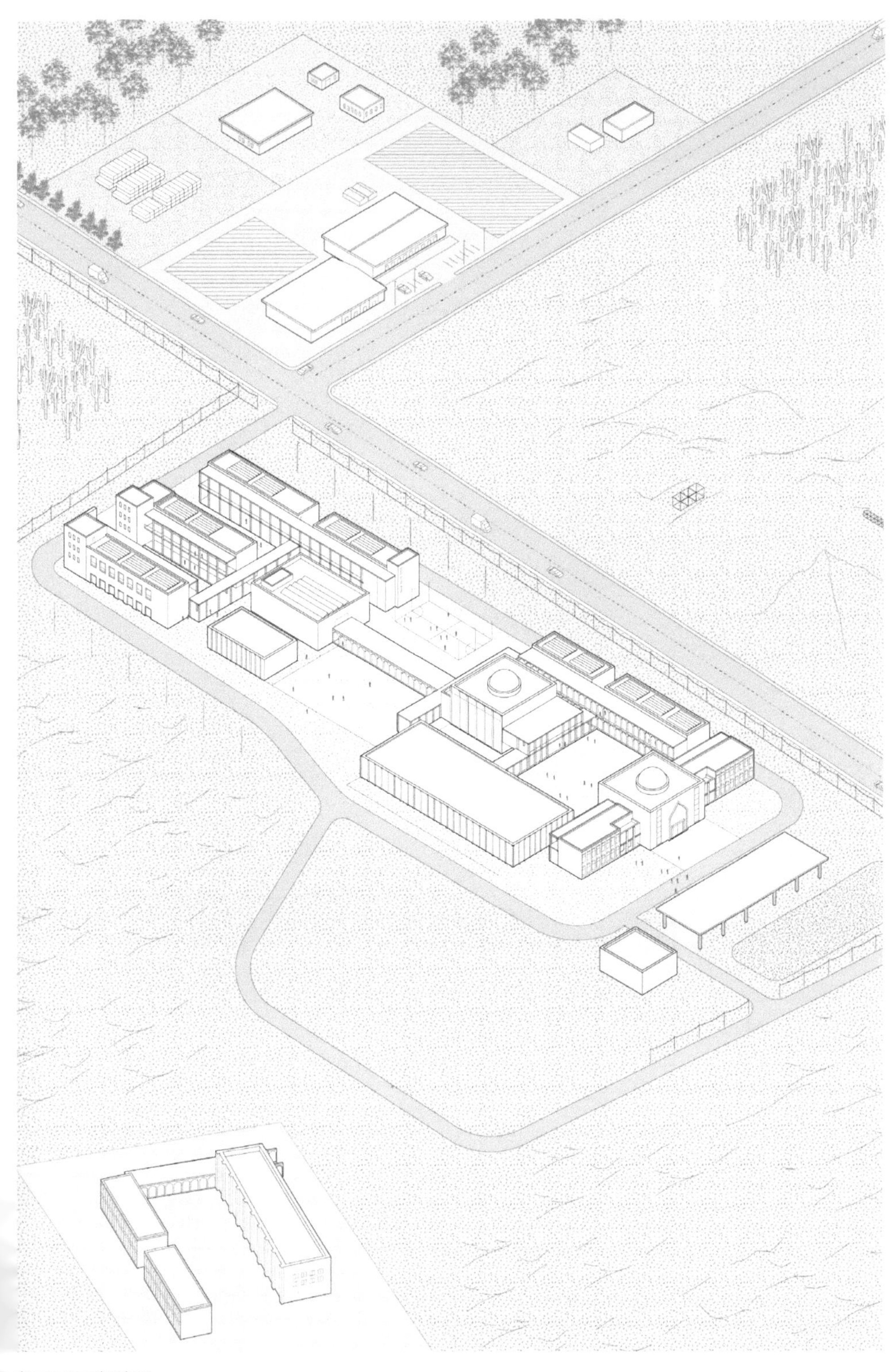

Axonometric view.

Pakistan-China Friendship Center

Location	Islamabad, Pakistan
Floor Area	40,500 square meters
Date	2012
Architect	Zhongyuan International Design Co., Ltd.
Client	Ministry of Culture of Pakistan

Arabic lattice-ornamented aluminum forming the external envelope of the building protects it from the sun.

The Pakistan-China Friendship Center in the heart of Islamabad, the cultural capital of Pakistan, is a monumental presence in one of the city's prime locations, characterized by green landscapes, museums, sports complexes, and several noteworthy institutional buildings. After a state visit by Chinese Premier Wen Jiabao to Pakistan in 2005, the center was founded as a welcoming business environment to expand cultural exchanges and set up a free-trade zone between the two countries. While China financed the project and ensured technical assistance, the Pakistani government was required to provide the almost 5 hectares of land for its development. After a contract notice was organized by the Chinese Ministry of Commerce together with the Pakistani Ministry of Culture, the Zhongyuan International Design Co., Ltd. (IPPR) was hired to design the building, while the Shanghai Construction Group was tasked with its construction. Both these Chinese state-owned enterprises had experience in foreign developments.

Programmatically speaking, the quite complex and articulated two-floor center includes three different auditoriums, collectively capable of hosting more than 3,000 people. Each auditorium is designed as a versatile and reconfigurable space, facilitating a variety of events, from conferences and concerts to discussions and presentations. There are also several restaurants that provide spaces for relaxation and dining. In addition, exhibition spaces carved out of the main circulation area can host cultural displays, trade fairs, and other events. The ground floor also has a mosque.

The formal composition of the building is quite clear and easily recognizable from its exterior. The whole complex is made up of three main volumes clad in local red bricks interrupted by glazed curtain walls and surrounded by a unifying envelope made of ornamented aluminum plates each measuring 1 × 1 meter; the plates were prefabricated and decorated using high-precision molding and high-pressure water cutting. The pattern of the ornamented lattice combines elements from traditional Chinese and Islamic architecture. It draws on both the Mashrabiya or "jali," the lattice feature in Islamic architecture, and on the "plum blossom," a typical element used in the windows of traditional Chinese houses. The building also includes central interior courtyards, with colonnades inspired by Chinese and Islamic gardens. The design by the IPPR integrates Chinese and Pakistani cultural features.

Territorial map on two scales.

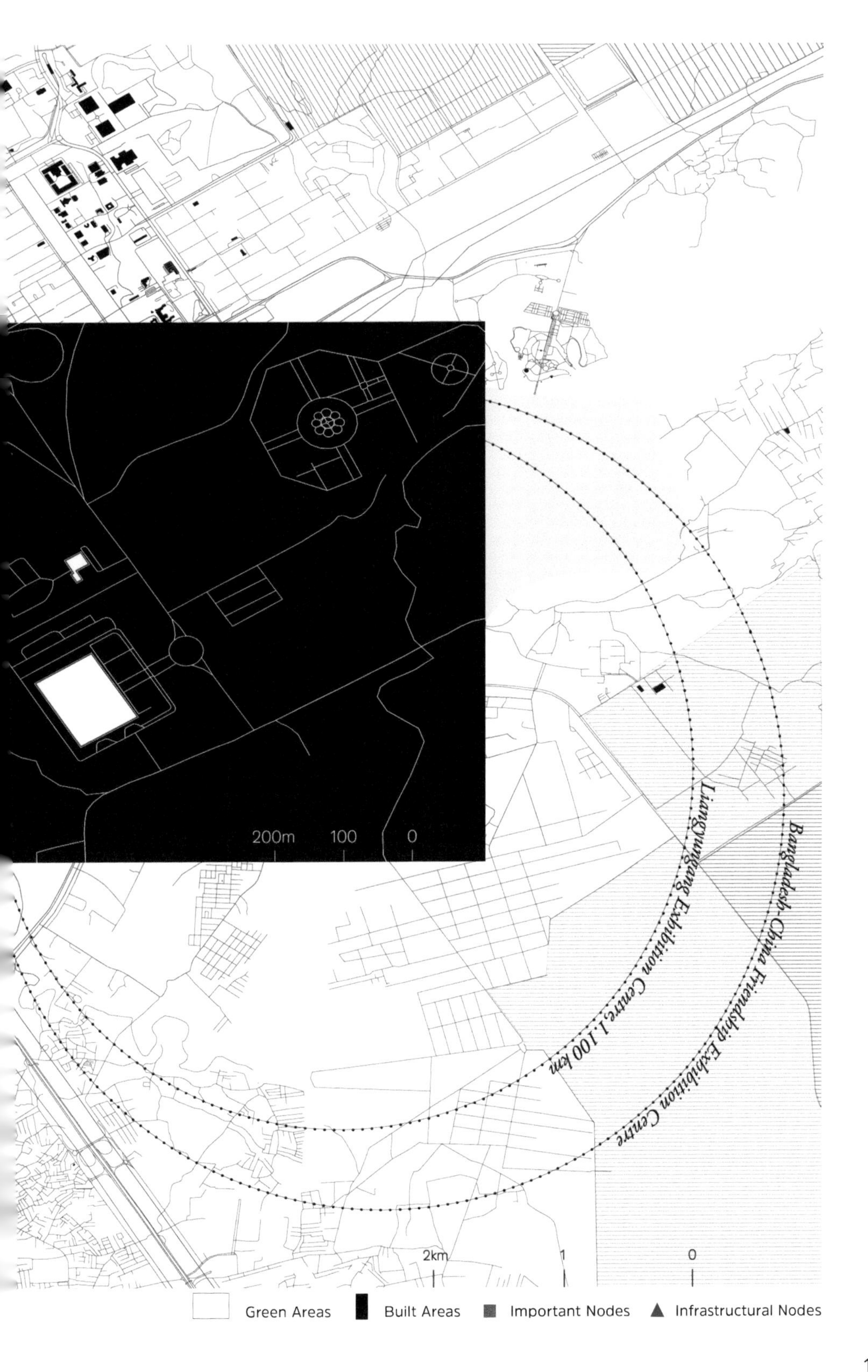
200m
100
0
Liangyungang Exhibition Centre, 1,100 km
Bangladesh-China Friendship Exhibition Centre
2km
1
0
Green Areas
Built Areas
Important Nodes
Infrastructural Nodes

Rounded arches inspired by Islamabad architecture

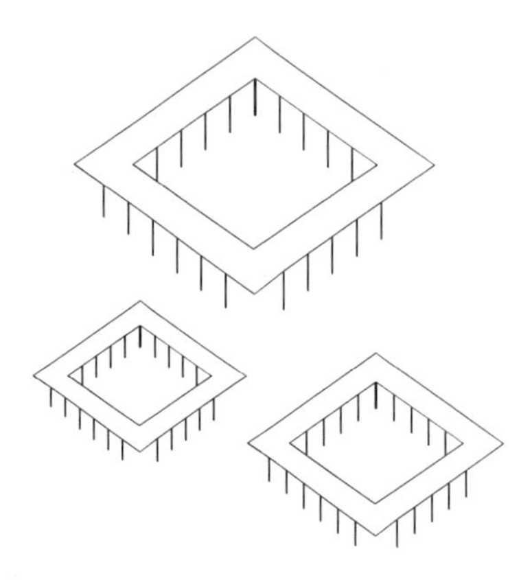

Courtyards surrounded by porches in Chinese style

Islamic-style spherical dome

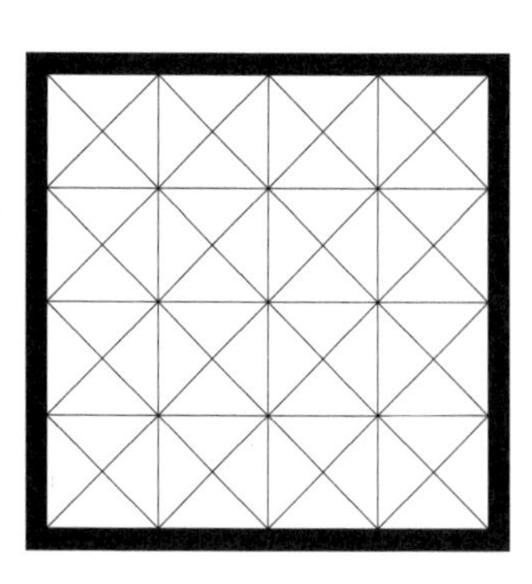

Façade following traditional techniques

Morphological diagram showing the main elements composing the architectural language.

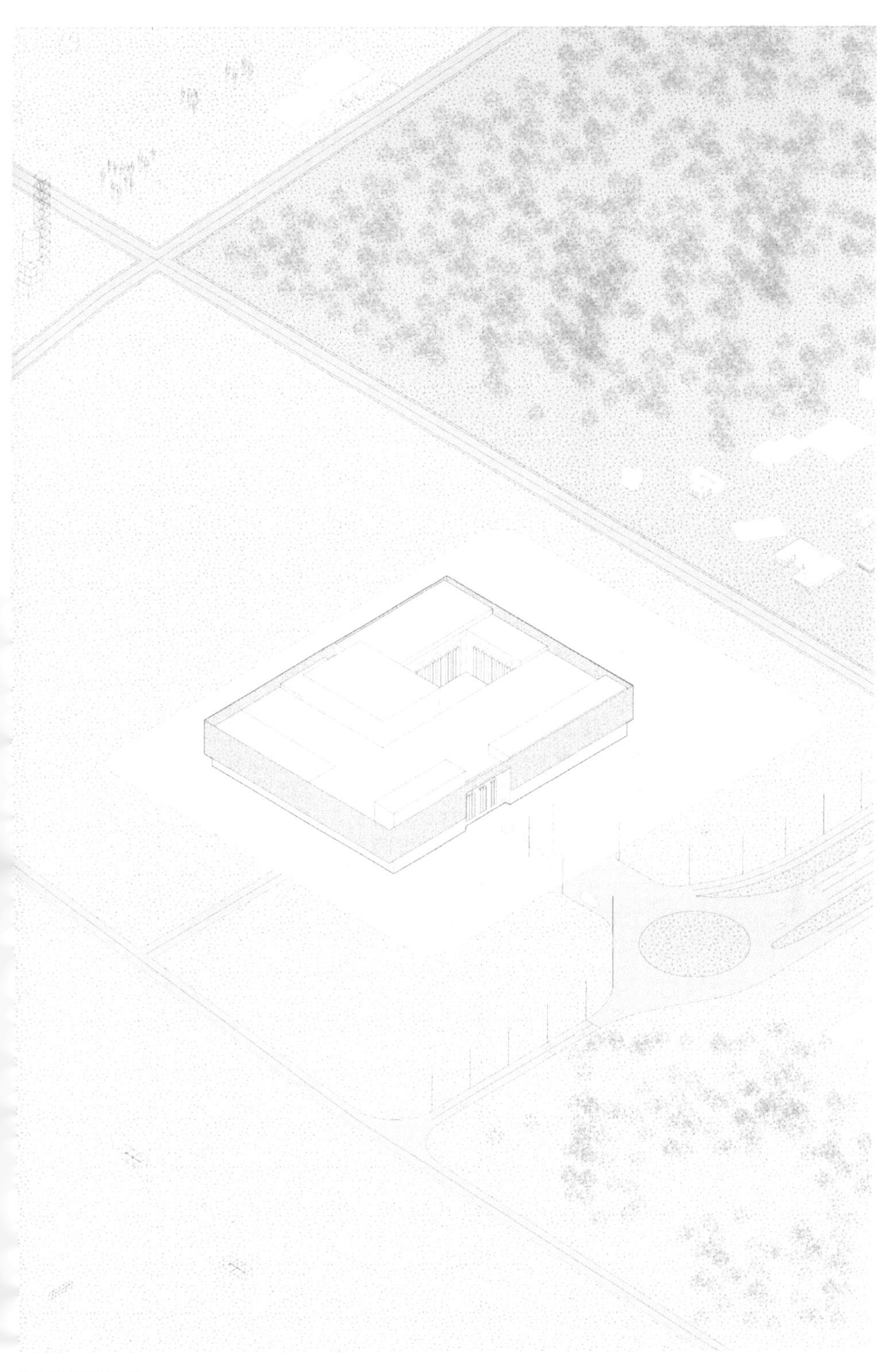

.xonometric view.

WORLDS OF SPECIAL RULES

Architecture between Humans and Data-Driven Machines

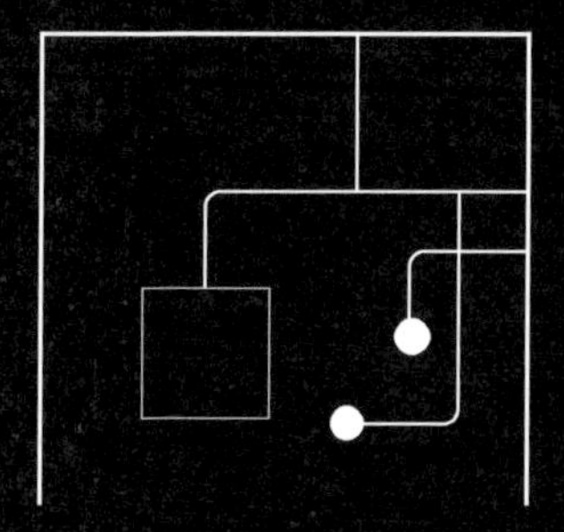

Lianglu-Cuntan Free-Trade Port Area

Location Chongqing, China
Built Area 3.88 square kilometers
Date 2020

Concrete oiers and automated steel cranes allow the movement of cargos between the river and the inland.

The Lianglu-Cuntan Free-Trade Port Area is situated at the strategic intersection of the New International Land-Sea Trade Corridor – one of many corridors in the Belt and Road Initiative established in 2019 and connecting China's Western regions (Chongqing, Guangxi, Guizhou, Gansu, Qinghai, Xinjiang, Yunnan, and Ningxia) with the Association of Southeast Asian Nations (ASEAN) – and the Yangtze River Economic Belt, an economic route that includes eleven provinces and municipalities along the Yangtze river, from Yunnan in the west to Shanghai in the east. The port area emerged as a vital logistical node for the China-Europe Railway Express (Chengdu-Chongqing) running through Central Asia and the New Western Land-Sea Corridor. Established in 2017, it is a pilot project to enhance Chongqing's strategic connection with Western countries and speed up the development of new trade. As China's largest inland river logistics port, it connects multi-modal transportation on canals, trains, and roads.

The whole area is divided into two separate but interconnected zones, i.e. the port and the manufacturing hub. The port features an 800-meter-long multi-story concrete platform suspended over the water. This space allows the processing of standard containers (12 meters long). Ball transfer units assist container hauling, the RFID and GPS tags guide automated robot vehicles while metal cranes lifting shipping containers from the river to the land move on a rail grid. The port also has state-of-the-art equipment, for example the CT-type H986 X-ray inspection machine for customs clearance, capable of examining 200 containers per hour using artificial intelligence. These technologies optimize container storage and expedite inspection processes, streamlining inbound/outbound operations on the concrete pier. The manufacturing hub, hosting several high-tech, automotive, and food-processing enterprises, is situated just to the side of the port in the inland area. It comprises small shelters, prefabricated boxes, extensive warehouses, and high-tech industrial facilities, as well as an office building, organized around a rigid grid of multi-lane roads. Goods and raw materials are processed or stored here while waiting to be moved to other locations.

The architecture of the Lianglu-Cuntan port prioritizes the aesthetics of machines, thus echoing Archigram's utopian vision of their "Walking City."

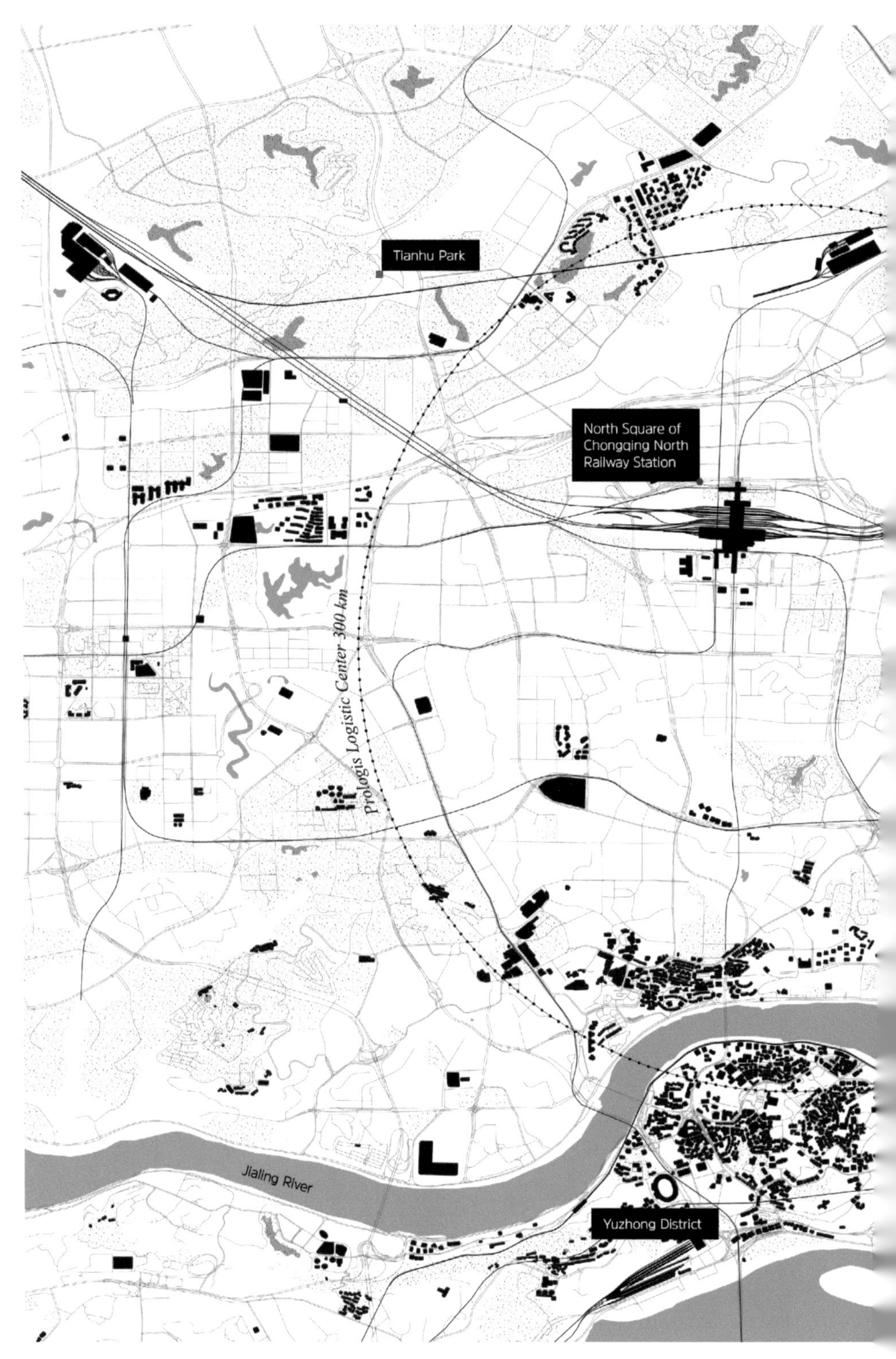

Territorial map on two scales.

Jiangbei District
0
0,5
1 km
200
100
Yangtze River
Chongqing Huang
Shang Scenic Area
River
0
1
2 km
Green Areas
Built Areas
Important Nodes
Infrastructural Nodes

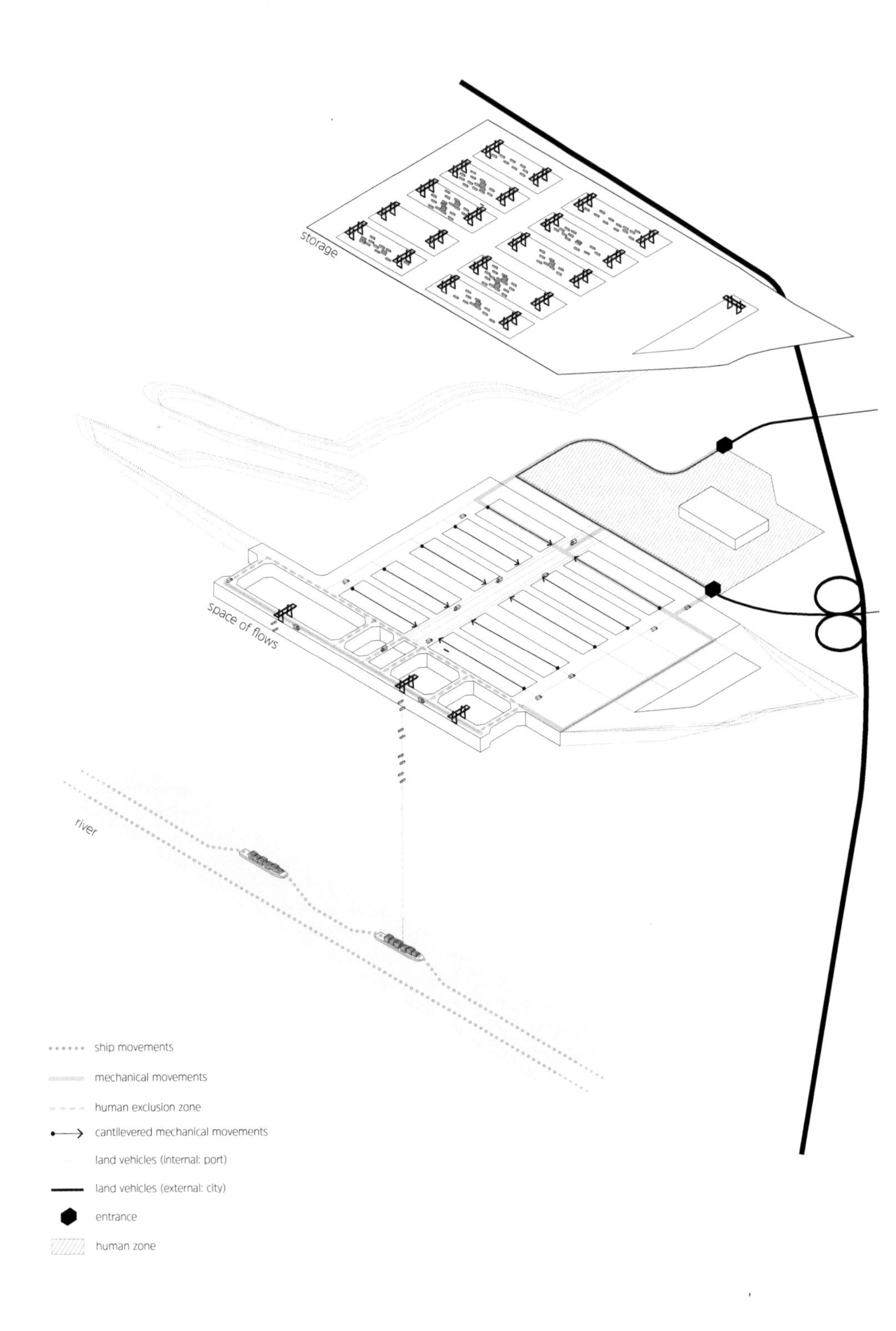

Circulation diagram showing the superimposition of human and non-human flows.

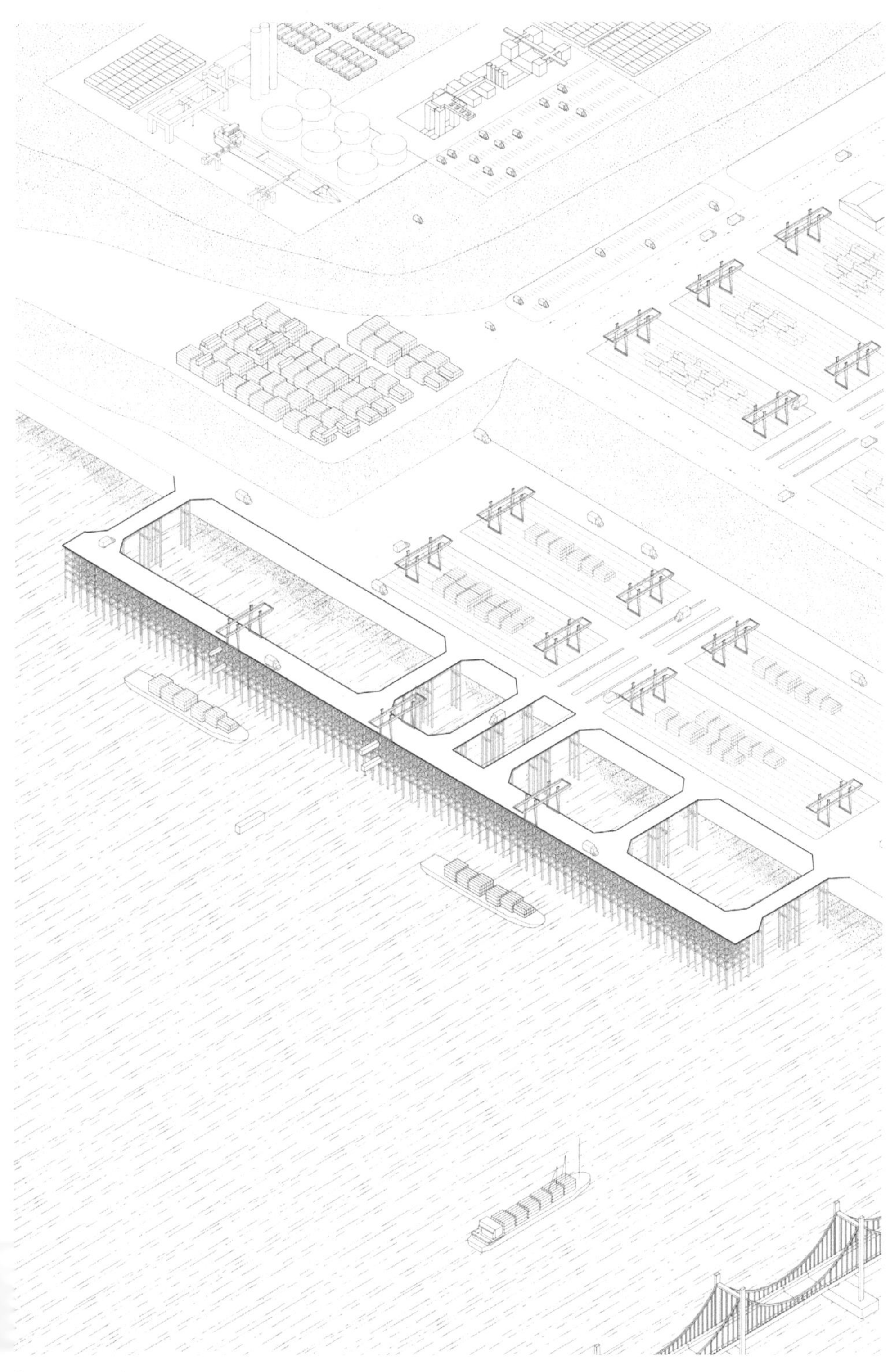

Axonometric view.

Prologis Logistics Center

Location Chongqing, China
Built Area 164,900 square meters
Date 2020
Client Prologis Co. Ltd.

Two-story warehouses are connected by white concrete ramps surrounding the buildings.

The Prologis Chongqing West Logistics Center is one of the many facilities built all over the world by the international logistics management company Prologis Co. Ltd. The Logistics Center is located in Chongqing Western New City, only 20 kilometers from the Chongqing downtown area and close to Highway G93, the main Chengdu-Chongqing intercity route. The center was intended to become an important logistic node of the BRI; it is also close (5.4 kilometers) to the Chongqing-Europe Railway departure station connecting Chongqing to the Asia-Pacific region and Europe.

The center was developed in two separate phases; it consists of four interconnected buildings with a total built area of 164,900 square meters. In order to maximize the land available in a relatively small plot, the buildings have two floors, making them very different from the more common warehouse type, usually just one-story high. Two white reinforced concrete ramps, winding around the buildings, allow trucks and other vehicles to access the second floor. The buildings are positioned around a main north-south central axis and three secondary axes in the opposite direction. On both floors, the main axis situated inbetween the buildings, acts as a platform for the loading/unloading of cargo. Most of the buildings' openings face toward the platform in the middle, while on the other side their appearance is solid and minimalistic.

The buildings' structure – fully exposed in the internal spaces – is made of prefabricated concrete columns and slabs with loading beams in two directions. The floor is designed as a polished concrete surface free from any obstruction so that it can be adapted to different uses. The simple and utilitarian envelope is made up of an independent steel structure covered by folded green and gray metal sheets – the colors of Prologis' brand. Small linear windows on the façade and skylights on the roof allow a minimum amount of natural light and air to enter the buildings.

Territorial map on two scales.

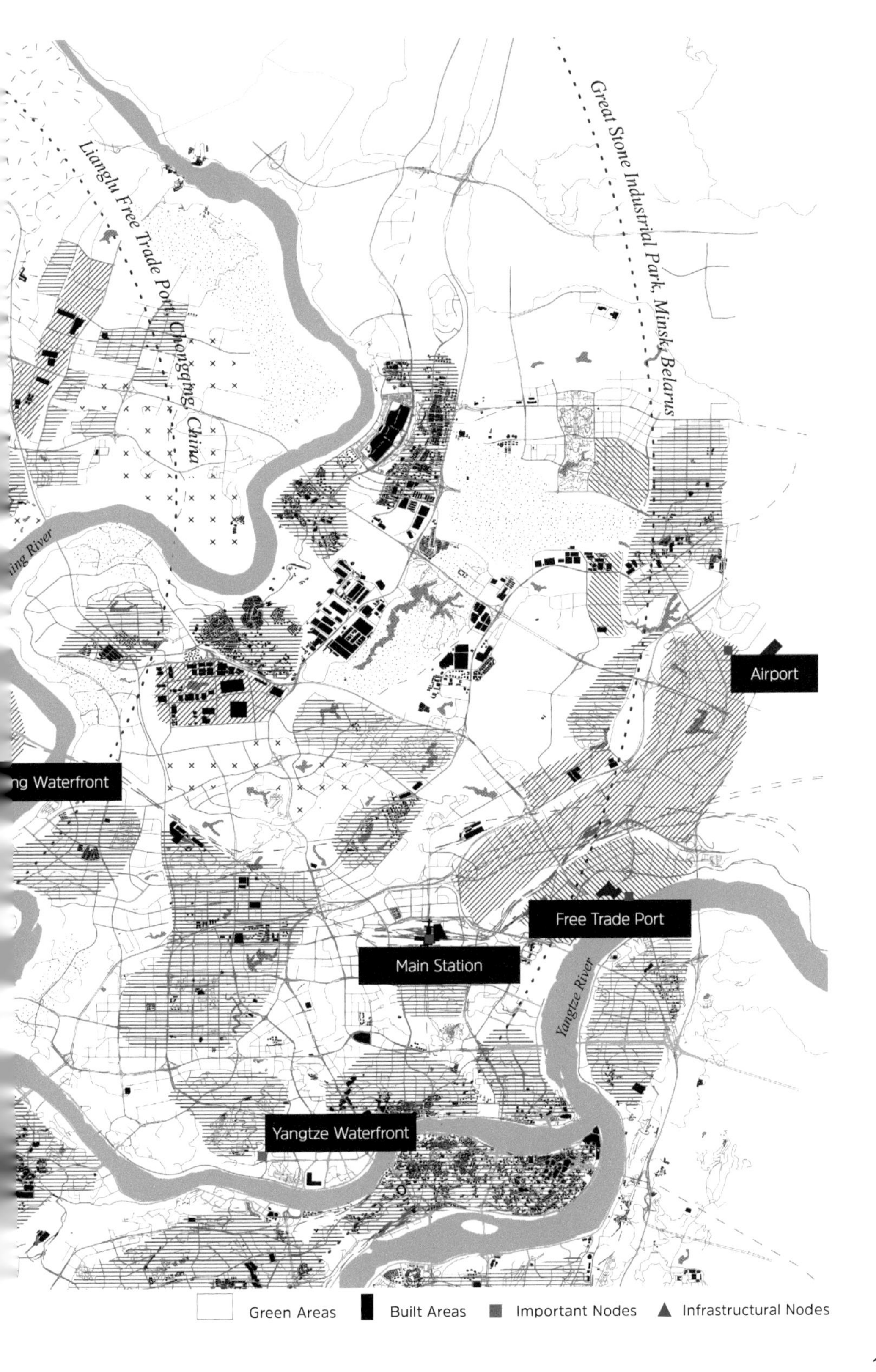

Prologis Logistics Center

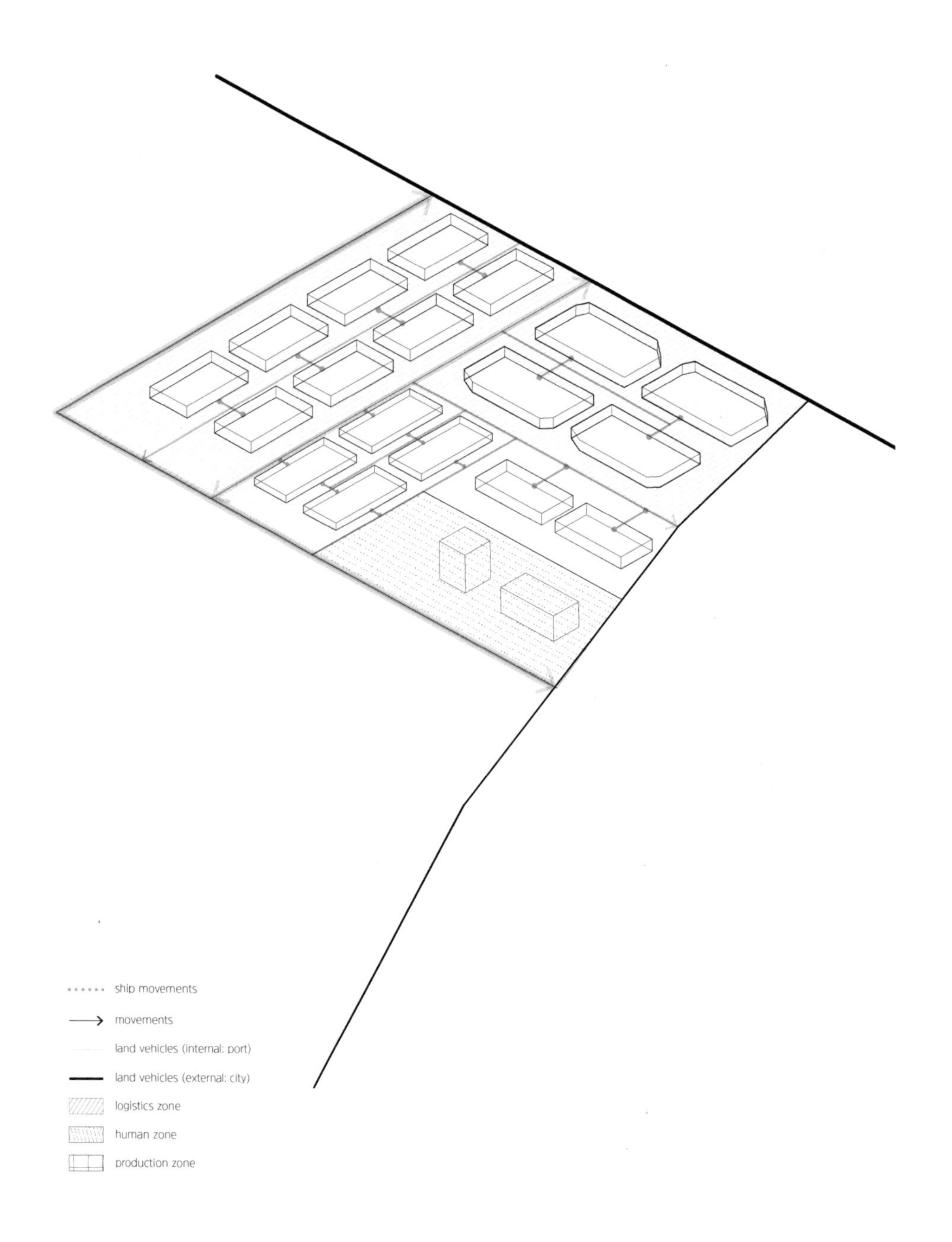

Circulation diagram showing the superimposition of human and non-human flows.

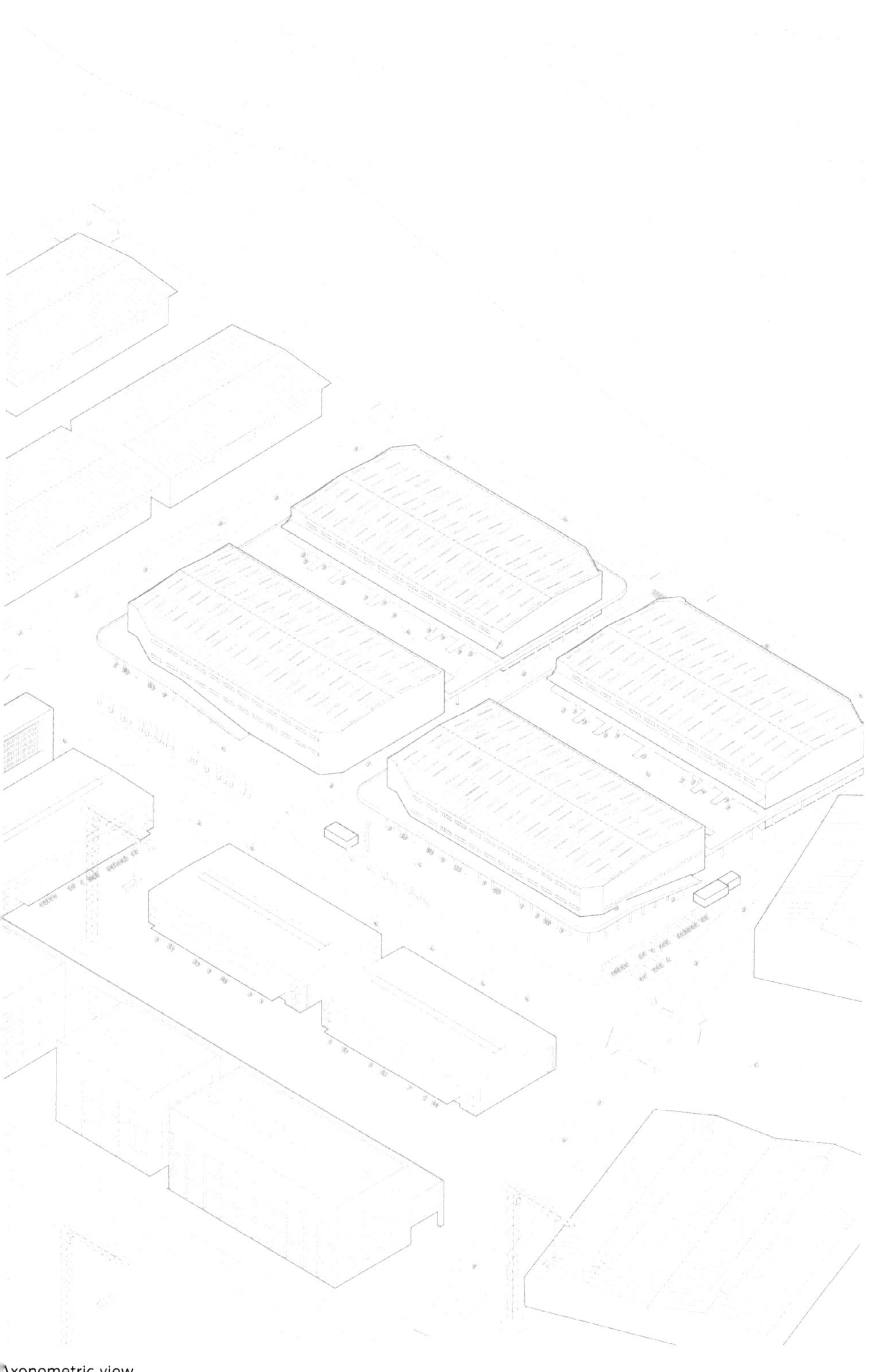

Axonometric view.

Khorgos Special Economic Zone

Location	Khorgos (KZ)/Horgos (CN)
Built Area	5.28 square kilometers
Date	2014 - ongoing
Architect	AECOM
Client	JSC International Center for Border Cooperation Khorgos

Foreign buyers and vendors entrance to the main shopping center, Khorgos.

The town of Khorgos nestles along the border between Kazakhstan and the Xinjiang Uyghur Autonomous Region in northwestern China; up until 2011 it was one of the most isolated and sparsely inhabited places on the planet. Located between the Kazakhstan Mountains and the Saryesik-Atyrau desert in Kazakhstan, Khorgos is only 130 kilometers from the enigmatic "Pole of Inaccessibility," the farthest point on the Eurasian continent from any sea or ocean. Strangely enough, one of the main reasons why Khorgos has become one of the most important and well-known nodes of the Belt and Road Initiative is due to a quite bizarre and accidental circumstance that may fascinate historians of technology: an 89-millimeter difference in width between the Chinese and Russian railway gauge systems triggered the need for a unique exchange hub: the dry port. As a result, Khorgos turned into a bustling commercial and business location soon after the launch of the BRI in Astana and the establishment of the Khorgos Special Economic Zone. The zone includes the dry port on one side and the International Center of Border Cooperation (ICBC) on the other. This latter development is based on a visionary urban master plan, designed by the renowned integrated design firm AECOM; the master plan envisages a mixed development program, with an emphasis on trade and tourism, straddling the borders of China and Kazakhstan.

However, only a few of the buildings and infrastructure that are part of this monumental plan have already been constructed on the Kazakh side of the border. The entrance building symbolizes the link between the two countries; the prefabricated construction, with its three-dimensional façade elements, provides access to the free-trade zone governed by its own regulations. The monumental gateway leads to a vast space organized around a central green axis. Here, the master plan by AECOM envisages a linear city hosting an international university, hotels, sanatoria, sports complexes and an ethno-park, in addition to trade and logistics facilities. Strangely enough, the vision for the new city lacks provisions for housing, shops, or other infrastructures necessary for urban life; this raises questions about how and if it can become a sustainable development model for the whole area. On the other hand, the Chinese side with its large shopping area of over 5 square kilometers, is already completely functioning and animated. Its shopping centers with long galleries and sales booths host a continuous flow of people from different parts of the world. In fact, since the ICBC straddles the border, it is neutral territory. No visas are required to cross the border here and several thousand shoppers come every day from Kazakhstan and nearby Central Asian countries to buy products at affordable prices. Yet, despite the grand vision that ICBC might become a new Dubai, its completion is still quite unclear.

On the other side, the Khorgos Special Economic Zone consists chiefly of the dry port: a highly automated environment, teeming with East-West freight and product flows, efficiently tracked by GPS and shorter-range wireless tracking tools. Here, freight and shipments are organized in advance, generating a continuous and simultaneous flow of goods that intersects with human activity in only a few dedicated zones established for inspection and exchange. More than ten years after its launch, the Khorgos Free Zone remains a fragmented intervention. While the automated dry port thrives, the humanized ICBC languishes.

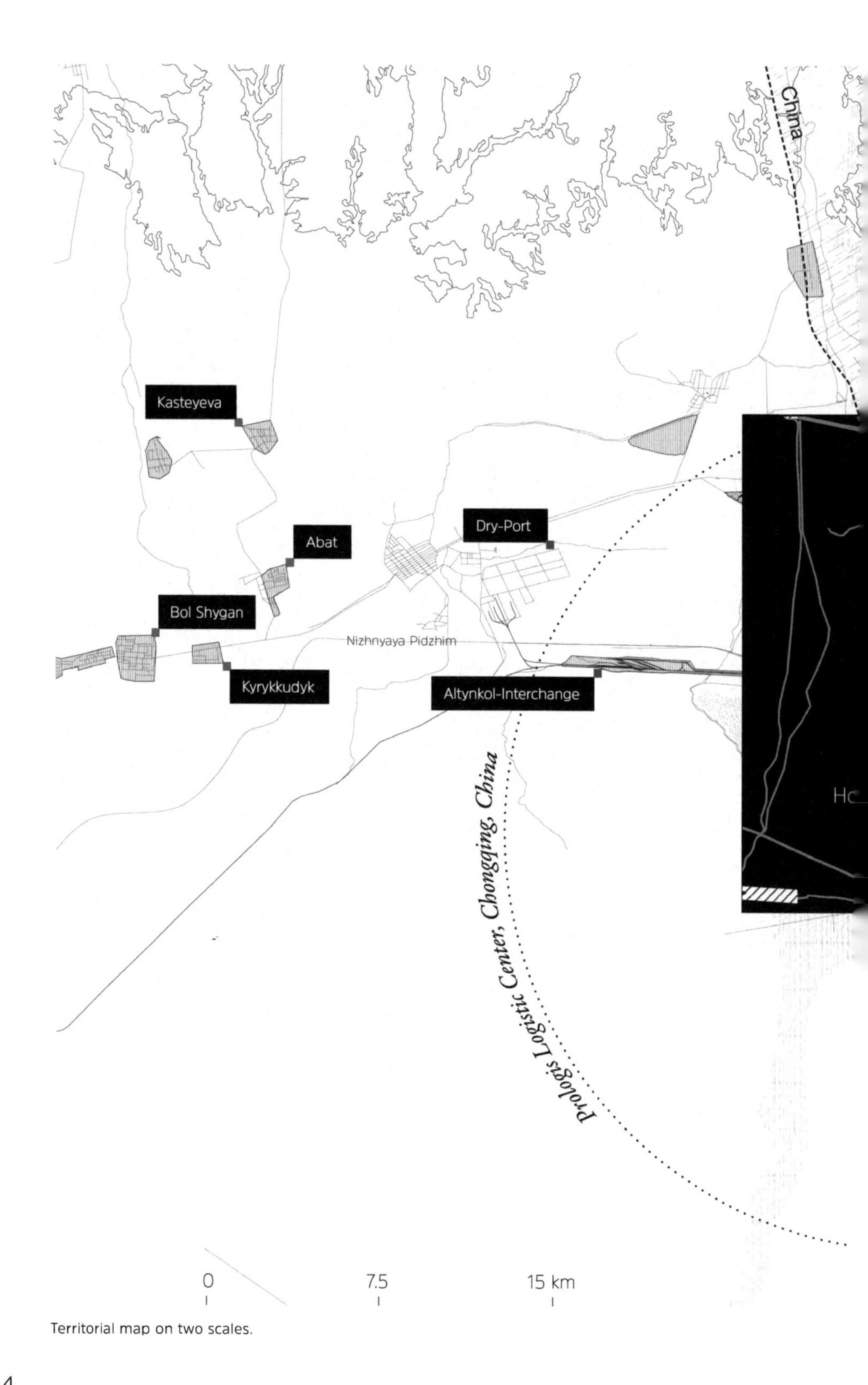

Territorial map on two scales.

Nongtianer Village
Chengxisancun
Ke Kedala Village
Green Areas
Built Areas
Important Nodes
Infrastructural Nodes

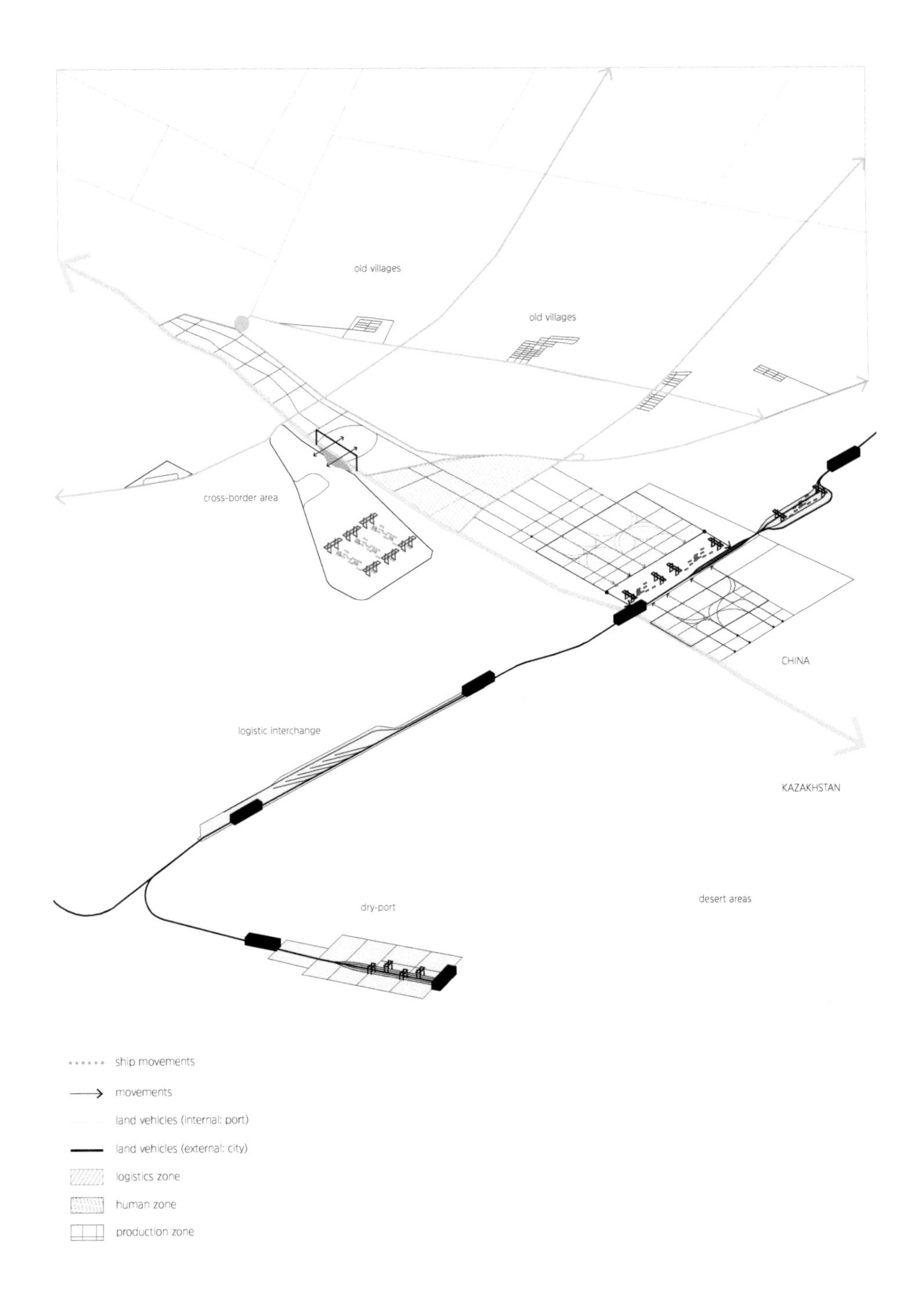

Circulation diagram showing the superimposition of human and non-human flows.

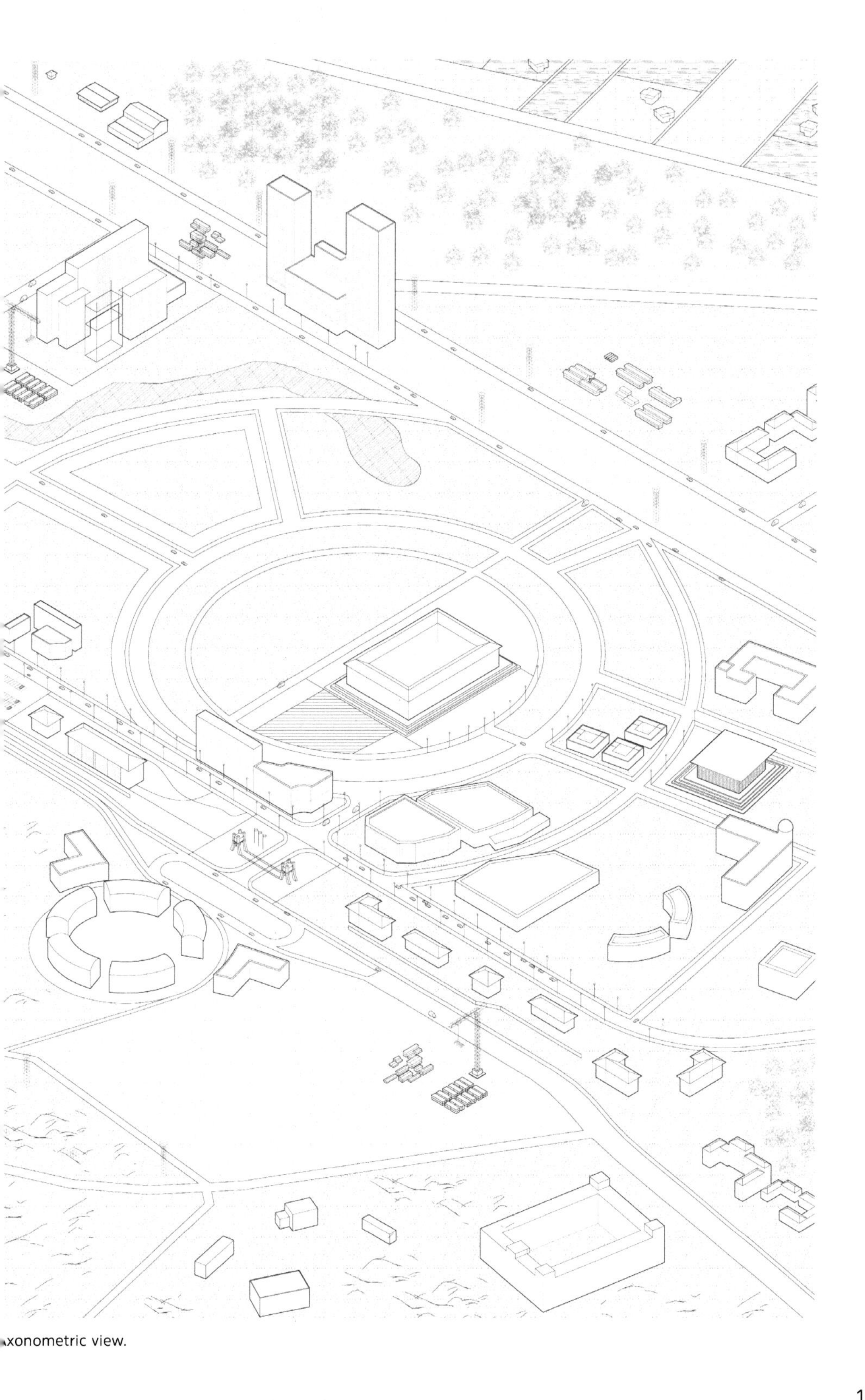

xonometric view.

Great Stone Industrial Park

Location	Minsk, Belarus
Built Area	8.5 square kilometers
Date	2015 - ongoing
Client	Great Stone Industrial Park Development Company

Multi-lane roads connect research centers and warehouses for the storage and processing of goods.

The Great Stone Industrial Park is a dynamic light manufacturing hub strategically positioned to provide tariff-free access to the Eurasian market, but close to the European Union. Located in the vicinity of Belarus's capital, Minsk, the park occupies a key position along the Northern Corridor of the New Silk Road trade route, one of the most critical in the current geopolitical situation.

The construction of the park is divided into five stages; the final phase is expected to be completed by 2060 and accommodate over 130,000 residents, as outlined in the official planning and promotional documents. Within this enormous development project, the park offers a range of ready-to-use manufacturing facilities that can be either rented or purchased, thus enabling a quick production setup. Alternatively, business enterprises have the option to build custom-made production facilities within the park. The overall plan revolves around a mix of four main functional zones: residential buildings, public buildings, industrial buildings, and extensive landscaping with recreational zones.

The architecture of the Great Stone Industrial Park reflects the distinctive features usually associated with an industrial park typology. From a formal point of view, there are no high-rise buildings or spectacular architectural designs, and most buildings are limited to two or three stories. Specific, variegated architectural features, such as polished brick cladding, sloping roofs, shiny glazed façades and colored envelopes, are prioritized only for specific buildings, such as the research and development laboratories and exhibition or conference centers; instead, all the other buildings, made of prefabricated steel structures wrapped around insulated concrete slabs, speak the language of global logistics architecture. Although the vision of the park is decisively business oriented, its multi-stage plan deliberately conjures up the image of a city, with an emphasis on places for people and strong infrastructural connections, such as a multi-lane road transportation network that links the different functional areas using several means of transportation, including cars, trucks, and even dedicated buses for workers and their families. Six- to eight-lane roads are built based on codified Chinese practices for infrastructure development. Street sections feature flower beds that separate access lines; wider roads near crossroads make the regulation of vehicular traffic more efficient. The big grid pattern, based on the Chinese industrial parks model, generates large plots measuring 300 to 500 meters by 300 to 500 meters; this results in a cellular urban environment where the essential aspects of urban life – namely housing, working, and consuming – are internalized within each block.

The Great Stone Industrial Park merges logistics architecture with conference centers, offices, commercial shops, multi-story housing buildings, and service facilities. Spaces dedicated to goods and humans are thus present and overlap within the same urban grid. The park is not merely a place for cargo to transit from one location to the other; on the contrary, it is a destination where goods come to be transformed and people arrive to establish a lasting presence.

Territorial map on two scales.

Mostishche
Minsk Airport
Green Areas
Built Areas
Important Nodes
Infrastructural Nodes

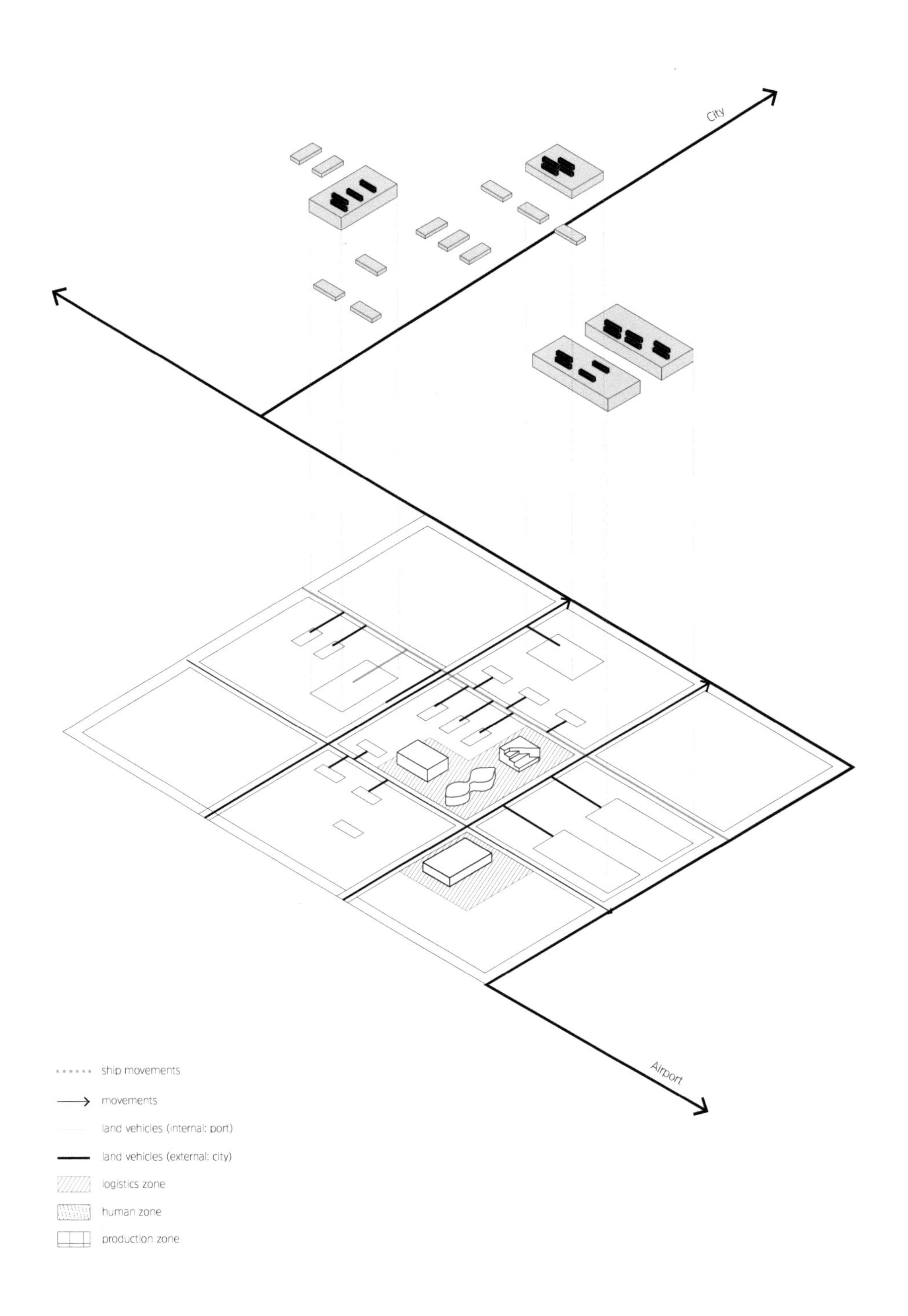

Circulation diagram showing the superimposition of human and non-human flows.

Axonometric view.

Suez Economic and Trade Cooperation Zone

Location	Suez, Egypt
Built Area	7.4 square kilometers
Date	2008 – ongoing
Client	Tianjin Economic-Technological Development Area Co. Ltd.

Leisure amenities in vernacular style are superimposed on a large manufacturing hub.

The Suez Economic and Trade Cooperation Zone, located at the crossroads between the Mediterranean and the Middle East, is a special economic zone operated by the Tianjin Economic-Technological Development Area (TEDA) Co. Ltd. The zone lies just 120 kilometers from Cairo and 40 kilometers from the city of Suez; it plays a key role in the program of the Belt and Road Initiative due to its proximity to the Suez Canal and the latter's strategic position between the Indian Ocean and the Mediterranean Sea. While plans for the zone had already been announced in 2008, its development gained momentum within the framework of the BRI and after President Xi Jinping's visit to Egypt in 2016.

The layout of the zone is designed as an orthogonal grid occupied by several industrial areas. Wide multi-lane roads are used in an orderly manner by trucks and cargo, while a network of underground infrastructures provide services in line with the standards of most Chinese industrial parks. The services include: electricity, roads, tap water, telecommunications, rainwater drainage, heating, gas, domestic sewage drainage, cable TV, firefighting facilities, and the preparation of the site for the future development of manufacturing buildings. A fiberglass manufacturer and high-voltage electric equipment workshop are, for example, among the production facilities available. However, unlike industrial facilities, the zone includes a series of supporting amenities, such as offices, a hotel, a conference center, and even an amusement park. The industrial warehouses are separated from the rest of the buildings by high walls, thus clearly dividing the zone into two separate but interconnected spaces: on the one hand an operational landscape and, on the other, a leisure space. This latter area is usually referred to as the "resort" in promotional materials. Visitors and businesses are considered "members," coexisting in a unique blend of small-scale vernacular buildings and modern offices surrounded by lush vegetation. As Keller Easterling noted in her seminal work *Extrastatecraft*, the presence of transient workers, businessmen, and tourists creates a temporary population contributing to the zone's vibrant business environment.

Interestingly enough, the dual nature of the zone is reflected by its architectural outlook. Indeed, while the industrial areas feature typical warehouses made of prefabricated concrete elements and with shining aluminum cladding on the roof, the other architectural elements are more in touch with the local context. This is clear, for instance, in the office buildings hosting the TEDA Special Economic Zone Management Company and the Egyptian-Chinese Joint Venture Company, both responsible for managing the area's development. These buildings are monumental; the big volumes are clad in local yellow stone, while the recessed windows accentuate the façade. The buildings' monumentality is further enhanced by large entrances characterized by a colonnade designed in a style vaguely reminiscent of Egyptian temples.

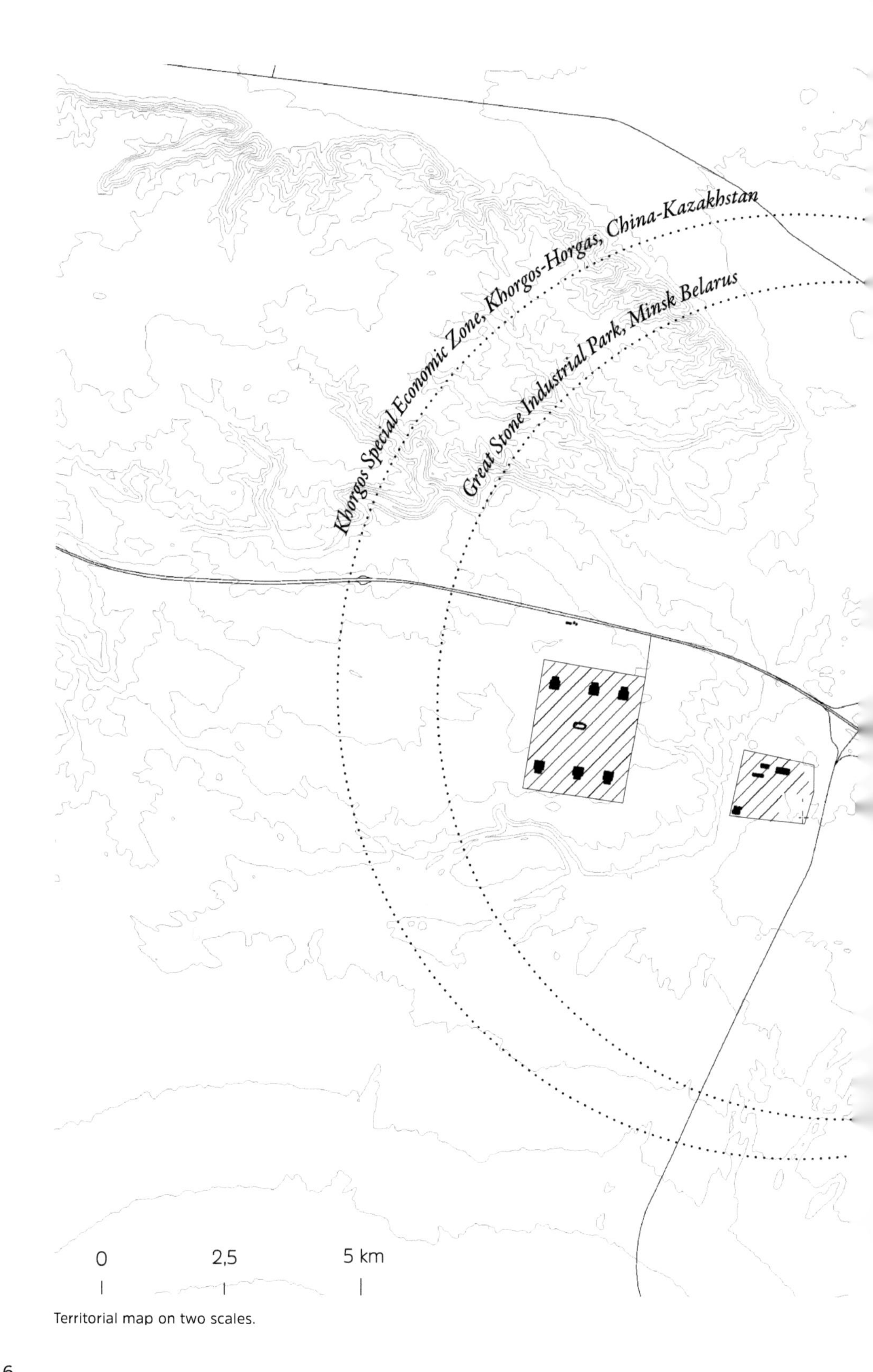

Territorial map on two scales.

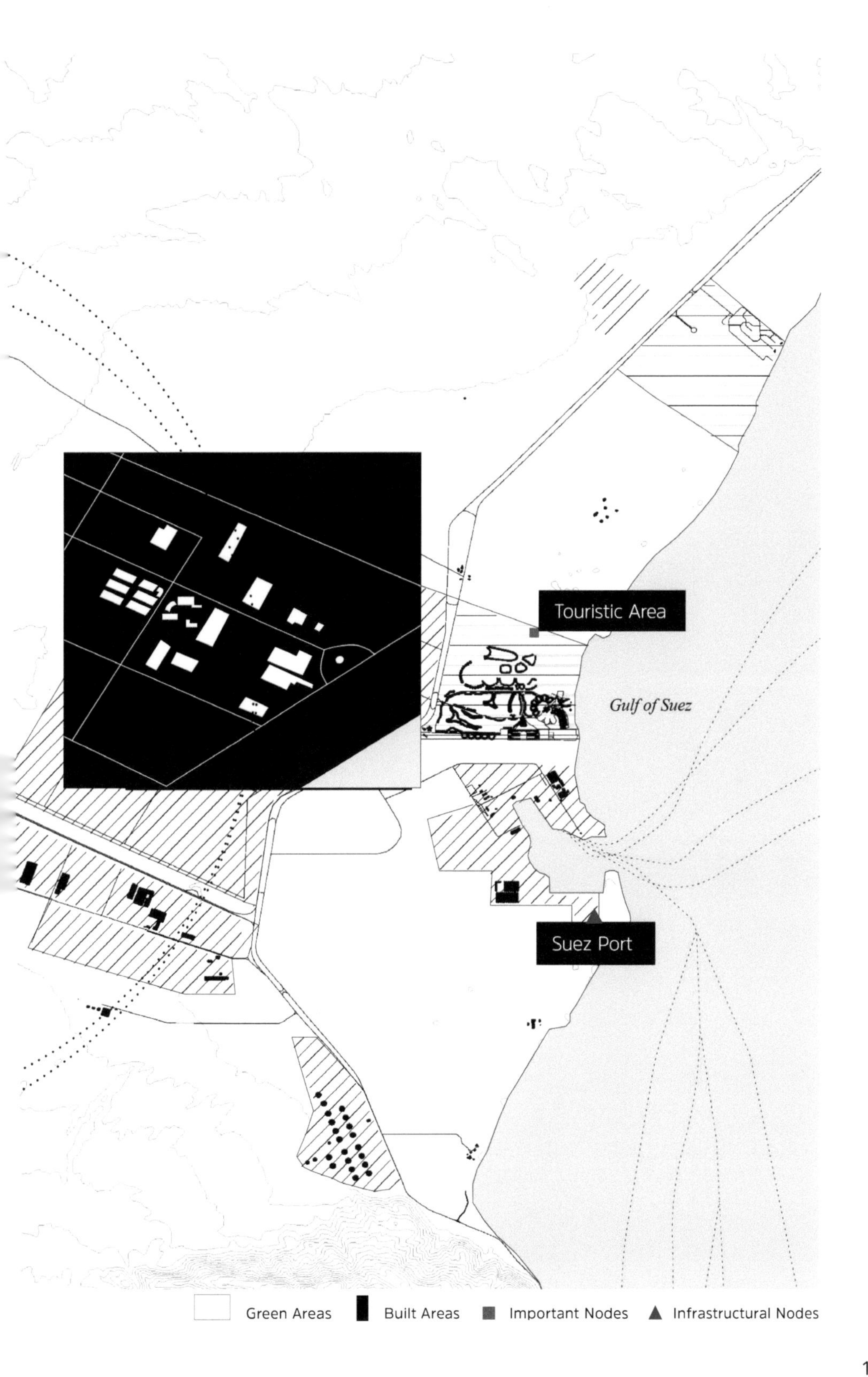
Touristic Area
Gulf of Suez
Suez Port
Green Areas
Built Areas
Important Nodes
Infrastructural Nodes

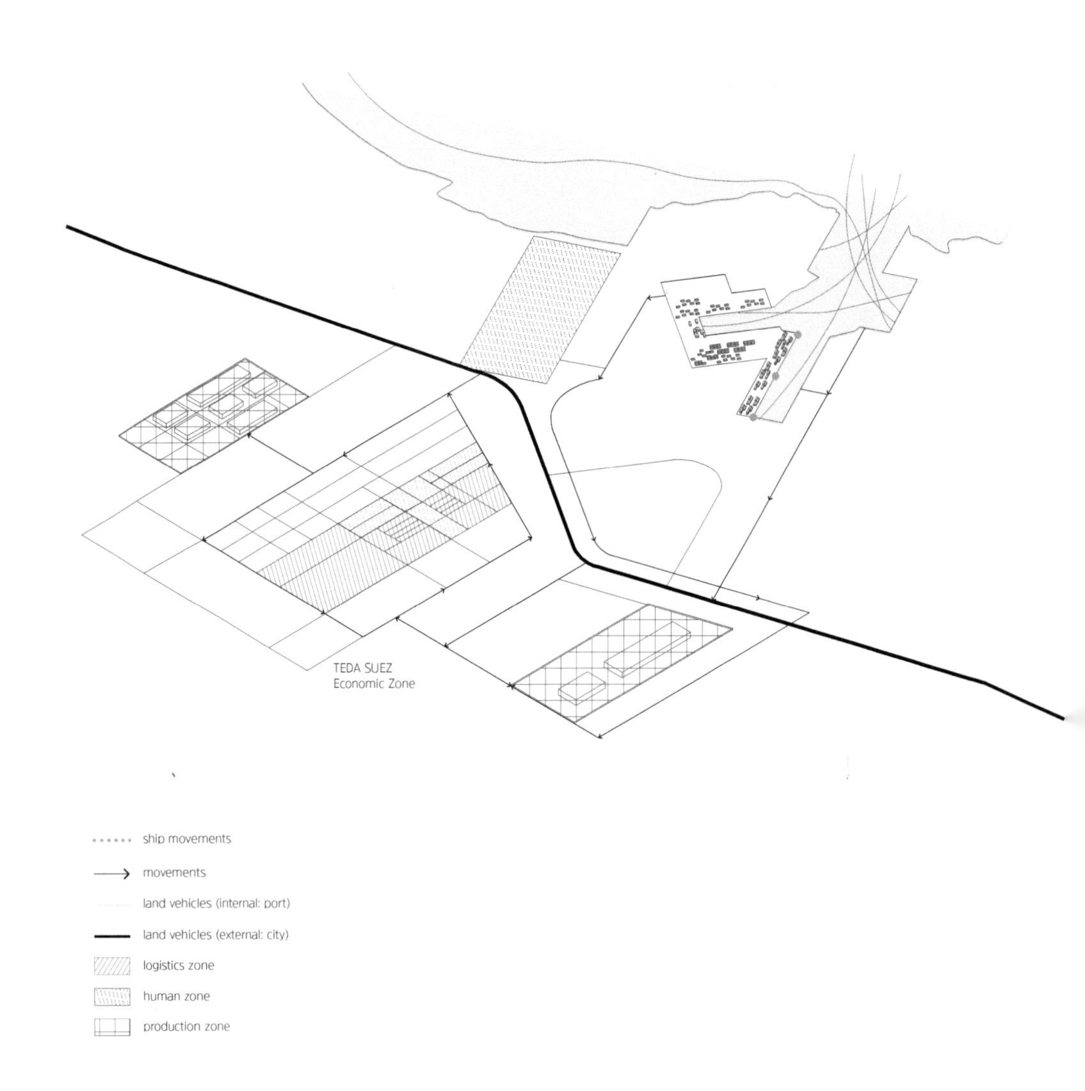

Circulation diagram showing the superimposition of human and non-human flows.

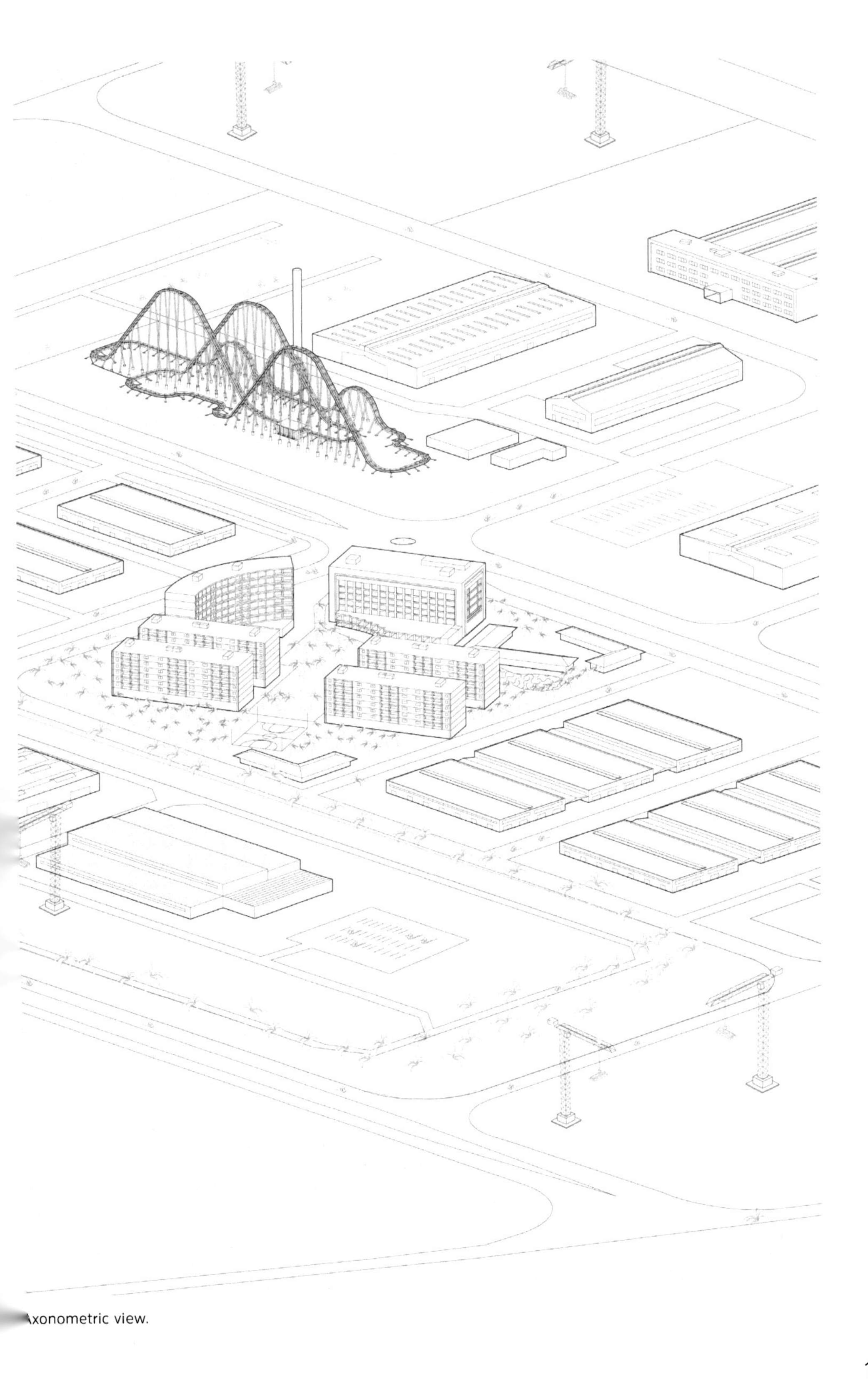

Axonometric view.

Mass Housing Enclaves

Between Standard Forms and Local Conditions

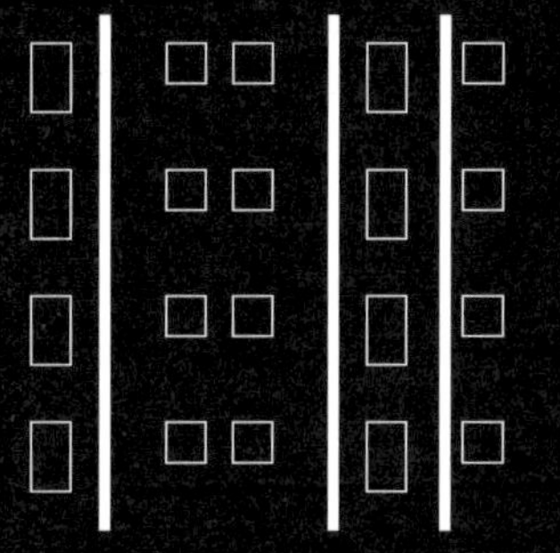

Ruiling Yayuan Housing Complex

Location Lanzhou, China
Floor Area 17,814 square meters
Date 2015
Client Lanzhou New Area Real Estate Development Co., Ltd.

High-rise residential buildings stand at the back of villas and communal facilities.

The ambitious Lanzhou New Area development project is located in China's arid and sandy northwest region, in Gansu province. It was launched in 2012 to boost trade interactions with the Central and western Asian countries participating in the BRI program, i.e., Kazakhstan, Uzbekistan, Pakistan, and several countries in the Balkan area. The development of the area located along the ancient Silk Road is part of an ambitious program launched in the early 21st century by the Chinese government; it aimed to promote the economic growth of twelve Chinese provinces in northwestern China which, until that time, had remained excluded from the fast-paced development of coastal areas. The new urban development – covering a total area of 25 square kilometers – includes an international airport, a modern high-speed rail station, industrial and logistics zones, a commercial district, an ecological forestry leisure zone, and a large residential zone for half a million new residents.

Seen from above, Ruiling Yayuan (one of the residential complexes of the Lanzhou New Area, completed in 2015 and developed by Lanzhou New Area Real Estate Development Co., Ltd.) is situated in a desertic landscape; it provides multi-story buildings that act as new homes for middle-class residents. The 4,824 housing units in the complex are present in both the high-rise buildings, set along the perimeter of the plot, and the mid-rise buildings in the middle. The complex's urban plan basically reflects the typical structure used by Chinese developers when organizing mixed-residential gated communities, especially after the "70/90 policy" was approved by the Chinese government in 2009. According to this regulation, 70% of the gross floor area of all new residential developments should host apartments of maximum 90 square meters, inserted in high-rise buildings for maximum profitability. On the other hand, mid- and low-rise buildings feature bigger apartments of around 165 square meters that can be sold as luxury products at a higher price.

The layout of the apartments also reflects a fairly consolidated formula of the Chinese real estate market: the entrance opens directly onto the living room, which typically also acts as a dining space. This includes a directly connected kitchen, often separated by a double sliding door to enhance natural ventilation and provide relative spatial flexibility. Every apartment includes a master bedroom, big bedrooms measuring around 25 square meters with an en suite bathroom. The apartments, and therefore the position of the windows, are always built in a north-south direction to enhance natural ventilation, so much so that the east and west walls are blind (often on both sides), in line with the customs of traditional Chinese residential architecture. Every apartment is endowed with an open space called *kong zhong hua yuan*, which we could translate as "courtyard" or "sky garden". This space is a recurring characteristic of almost any new urban apartment in China and is liberally exploited by real estate promoters. Indeed, while official documents report it as being an accessory space (contributing to the GFA with 50% of its area), it is often sold as an additional "room," and therefore as a supplementary surface area in the calculation of the value of the housing unit. The main element of differentiation is provided by the buildings' ornamental façades and the internal landscape made up of gracefully curved green islands, water pools, signage with foreign names, and lifestyle amenities, all deployed to conjure up a unique atmosphere.

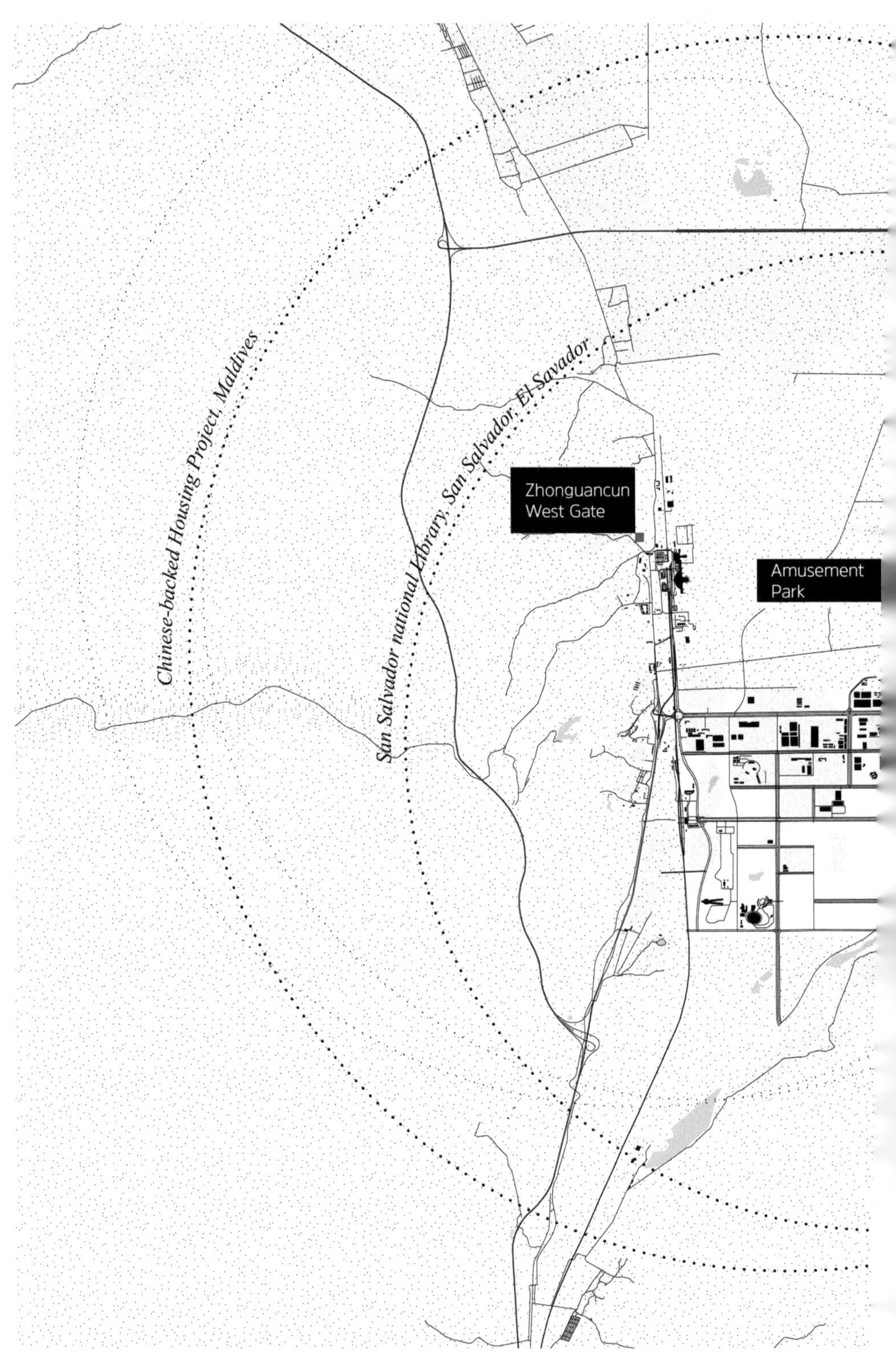

Territorial map on two scales.

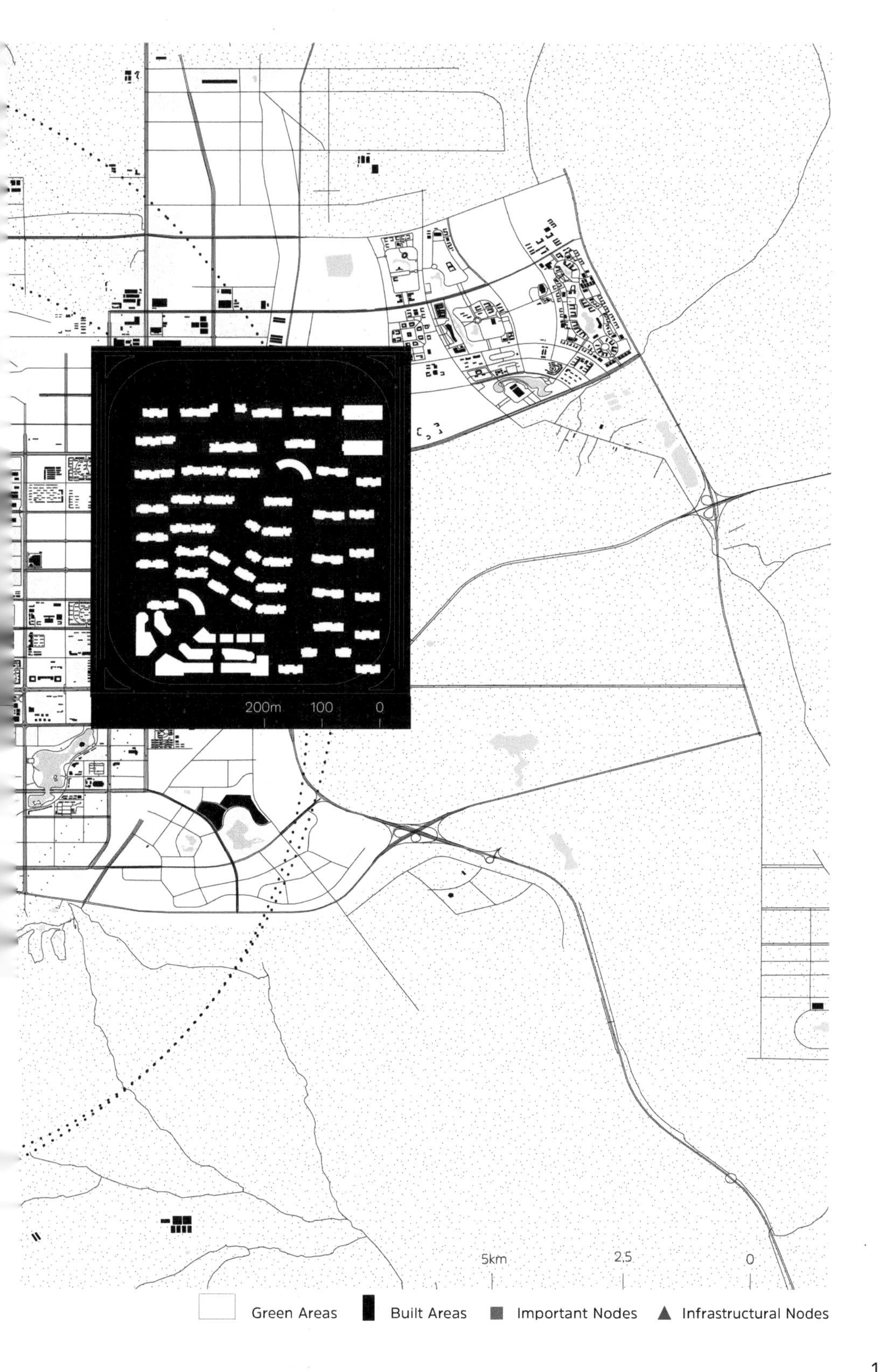
200m
100
0
5km
2.5
0
Green Areas
Built Areas
Important Nodes
Infrastructural Nodes

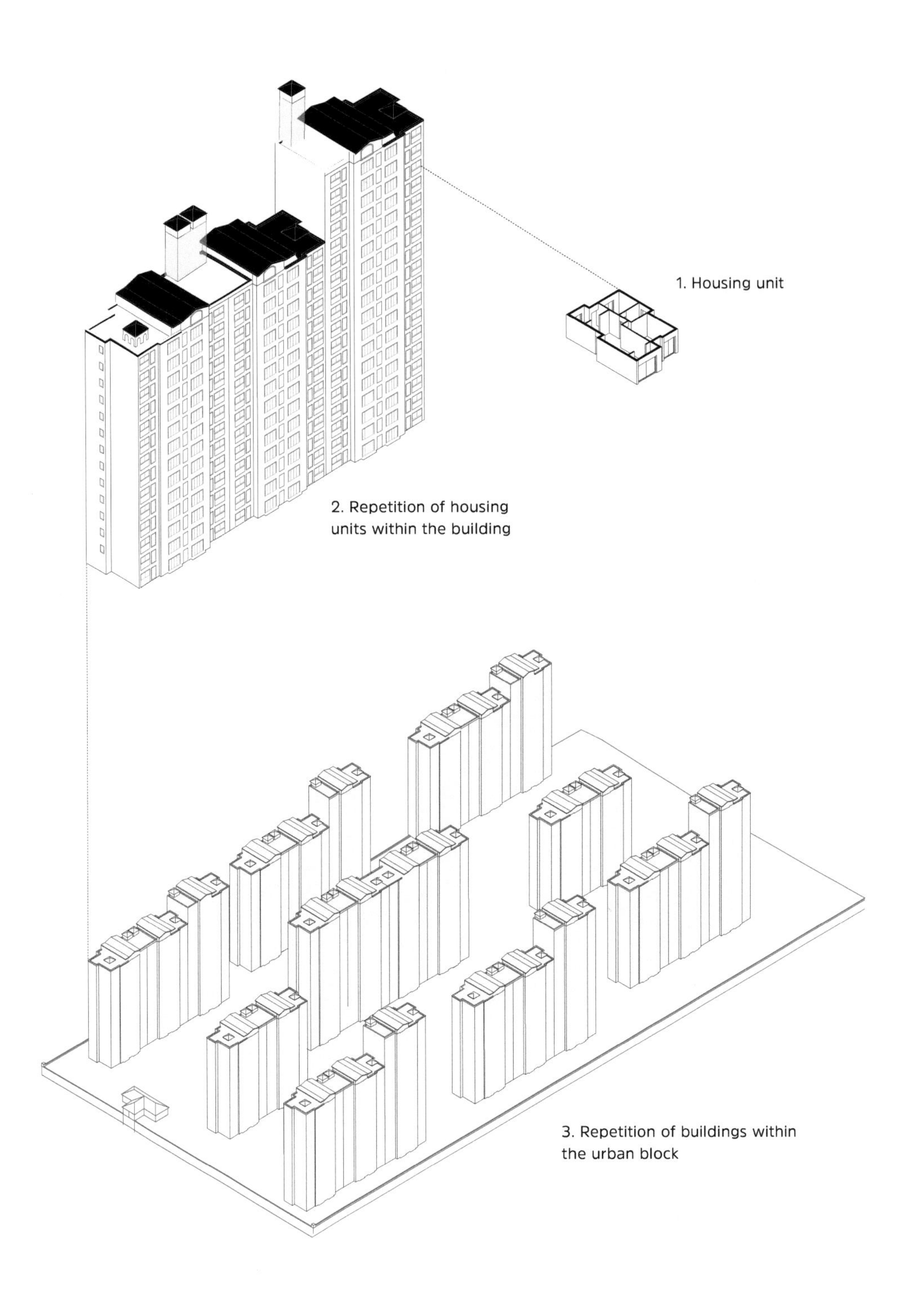

Morphological diagram showing the repetition of architectural elements at different scales.

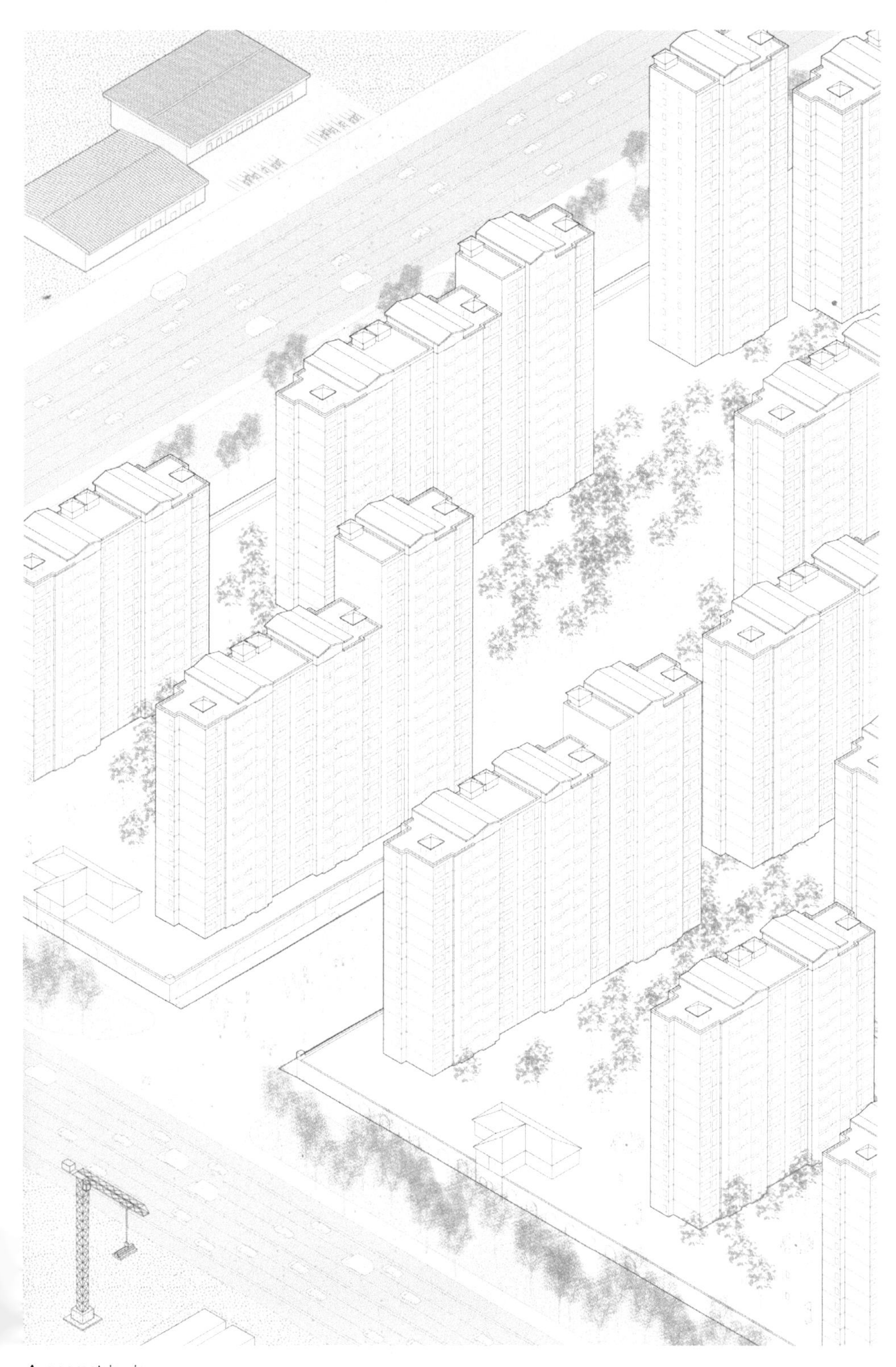

Axonometric view.

Hiyaa Housing Project

Location	Hulhumalé, Maldives
Floor Area	468,000 square meters
Date	2020
Architect	China's International Engineering Company
Client	Hulhumalé Development Unit

High-rise buildings are characterized by the repetition of modular elements and the variation of colors in their façades.

The Hiyaa Housing Project in Hulhumalé has been praised by official sources as a remarkable architectural initiative that addresses the pressing issue of the housing shortage and soaring rent prices in the Maldives. However, to understand the broader strategic importance of the project within the framework of the Belt and Road Initiative, it is crucial to go back to 2022, when the Maldives and China celebrated the fiftieth anniversary of their diplomatic relations. To mark this special occasion, the Chinese Embassy and the Ministry of Economic Development of the Maldives co-hosted a business forum focused on the joint implementation of the Belt and Road Initiative between the two countries. Three projects stand out among the many infrastructural investments that were implemented under this framework: the China-Maldives Friendship Bridge; the expansion of the Velana International Airport; and the Hiyaa Housing Project in Hulhumalé, the largest housing project in the Maldives.

The Hiyaa Housing Project covers an area of 468,000 square meters spread across nine islands in five atolls; it can accommodate nearly 30,000 residents in a total of 72 high-rise buildings developed in several phases by the China State Construction Engineering Corporation (CSCEC). The high-rise residential complex stands out against the island's skyline as a repetition of almost identical towers, at least when seen from the outside. Despite the overall appearance, the project features three different main types of residential buildings: square-based towers with one central distribution, rectangular-based towers with one central distribution, and rectangular-based towers with two access systems. This functional differentiation – reflecting a search for efficiency in the relationship between structural systems, building distribution, and variegated apartment layouts – is interpretable from the outside by different color strips on the building façades (red, blue, yellow, green, and purple). These colors serve to brighten up the rigid, severe, external concrete envelope characterizing the exterior appearance of this large mass of repetitive towers. Even though the construction technologies and spatial layouts are governed by a catalog of standardized layouts brought by the Chinese architects and developers, small adaptations to the local context are evident in several of the solutions. The open roof design of the central core of each tower, for instance, provides natural ventilation within the building – a way to combat the warm, humid environment of the Maldives. At the same time, the housing unit layouts are based on the dimensions and standards used in previous projects developed by CIEC in their homeland China; they are carefully planned with a minimum floor area of 51 square meters and have two bedrooms with en suite bathrooms in order to comply with local regulations imposed by the Hulhumalé Planning and Development Organization.

In other words, while "repetition, repetition, repetition" seems to be the mantra to describe projects such as the Hiyaa Housing development, differences thrive through continuous negotiation between the forces of globalization and small adjustments responding to local contingencies.

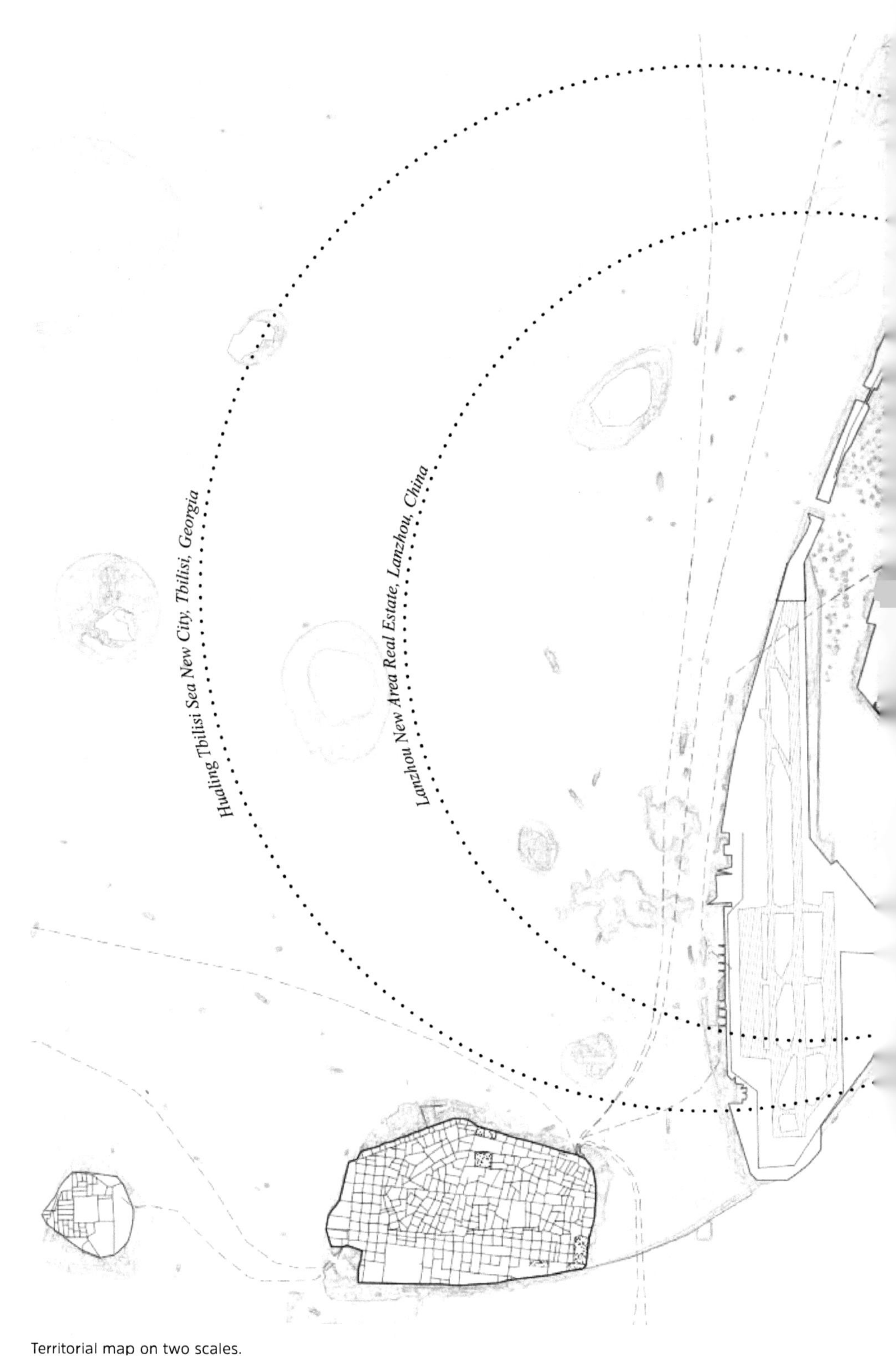

Territorial map on two scales.

5km
2,5
0
Green Areas
Built Areas
Important Nodes
Infrastructural Nodes

Hiyaa Housing Project

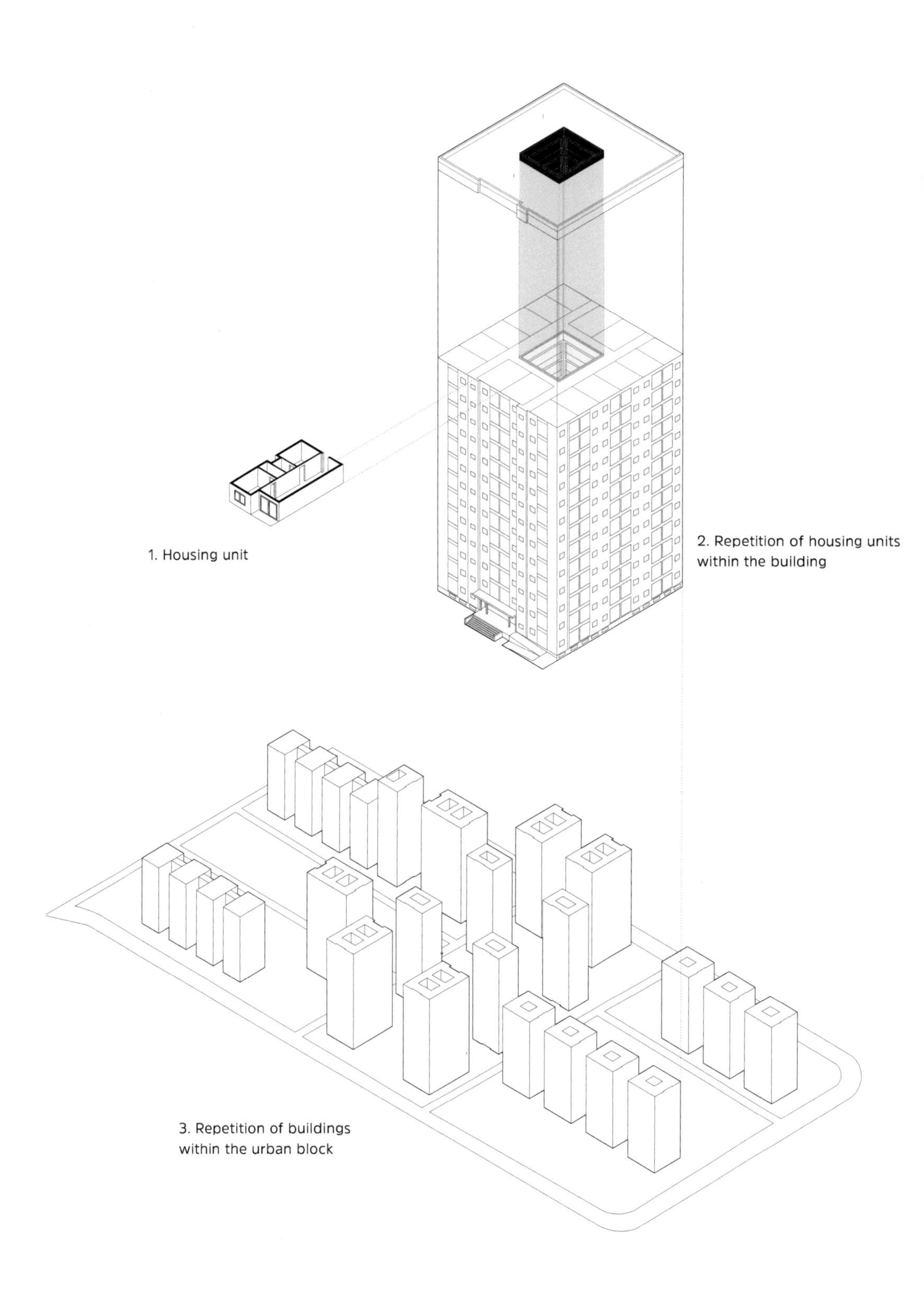

Morphological diagram showing the repetition of architectural elements at different scales.

Axonometric view.

Hualing Tbilisi Sea New City

Location Tbilisi, Georgia
Floor Area 4.2 square kilometers
Date 2008–2012 first phase; 2014–2022 second phase
Client Hualing Group

Residential buildings and communal spaces are made of an assemblage of Western-style structures and ornamentations.

Hualing Tbilisi Sea New City occupies a total area of 4.2 square kilometers, superimposing itself over a scattered group of small villages in the Bhal region of Georgia – a tidal flat extending approximately 15 kilometers inland from the coast of the Gulf of Khambhat. The project represents the flagship operation of the Hualing Group, a Chinese real estate developer. Founded in 1988 in Urumqi, China, the Hualing Group is a multi-property company managing over 30 enterprises with different business profiles and four wholesale markets with a combined space of three million square meters. After the company was provided with incentives by the Georgian government it began to invest in Georgia in 2007; since then it has implemented several major projects, including – in addition to Hualing Tbilisi Sea New City – the Hualing Tbilisi Sea Plaza, the Hualing Dormitory Hotel, the Hualing Free Industrial Park, the Hualing Hotel in Kutaisi, and the Hualing Wood Development, accounting for a total capital investment of around 500 million US dollars. As a result, the group is one of the biggest private investors in the country. The origins of Tbilisi Sea New City was the result of a deal between the Georgian government and the Hualing Group. In exchange for financing and building several temporary housing units for athletes to live in during the 2015 Youth Olympic Games – hosted that year by Georgia – the Chinese company was granted rights to develop 420 hectares of land for the new city.

Based on codified practices of new town developments in China, the final program of the Hualing Tblisi Sea New City consists of a large residential space divided into apartment and villa complexes surrounded by amenities and infrastructures, such as recreational zones, commercial spaces, a high school, college, library, exhibition hall, cinema, police station, administration, post office, clinics, gymnasium, and natatorium. The urban scheme is that of a Chinese gated community with fortress-like walls enclosing the residential buildings and their dedicated facilities. As in many other Chinese new city developments, a cozy but conventional atmosphere is intentionally generated by the medium-rise decorated buildings (nine and ten stories) with a landscaping consisting of water pools with curved green islands and extensive lawns. External building forms and ornaments are an assemblage of historicist details, including faux maroon bricks, ornately framed windows, and dozens of little circular balconies.

The size and layout of the apartments are highly standardized on the basis of previous developments by Hualing in China (68 square meters – two-room apartment with 1 bathroom; 78 square meters – three-room apartment with two bathrooms; 117–135 square meters – duplex five-room apartment with three bathrooms); this reflects the needs and expectations of middle-class homeowners who have emerged in China's post-reform era. However, while these housing types currently represent the norm in China, they appear somewhat novel in Georgia, providing a viable housing alternative for an emerging urban middle class.

In Georgia, Hualing Tbilisi Sea New City redefines the relationships between the territory and the capital by replicating China-made cities on greenfield sites.

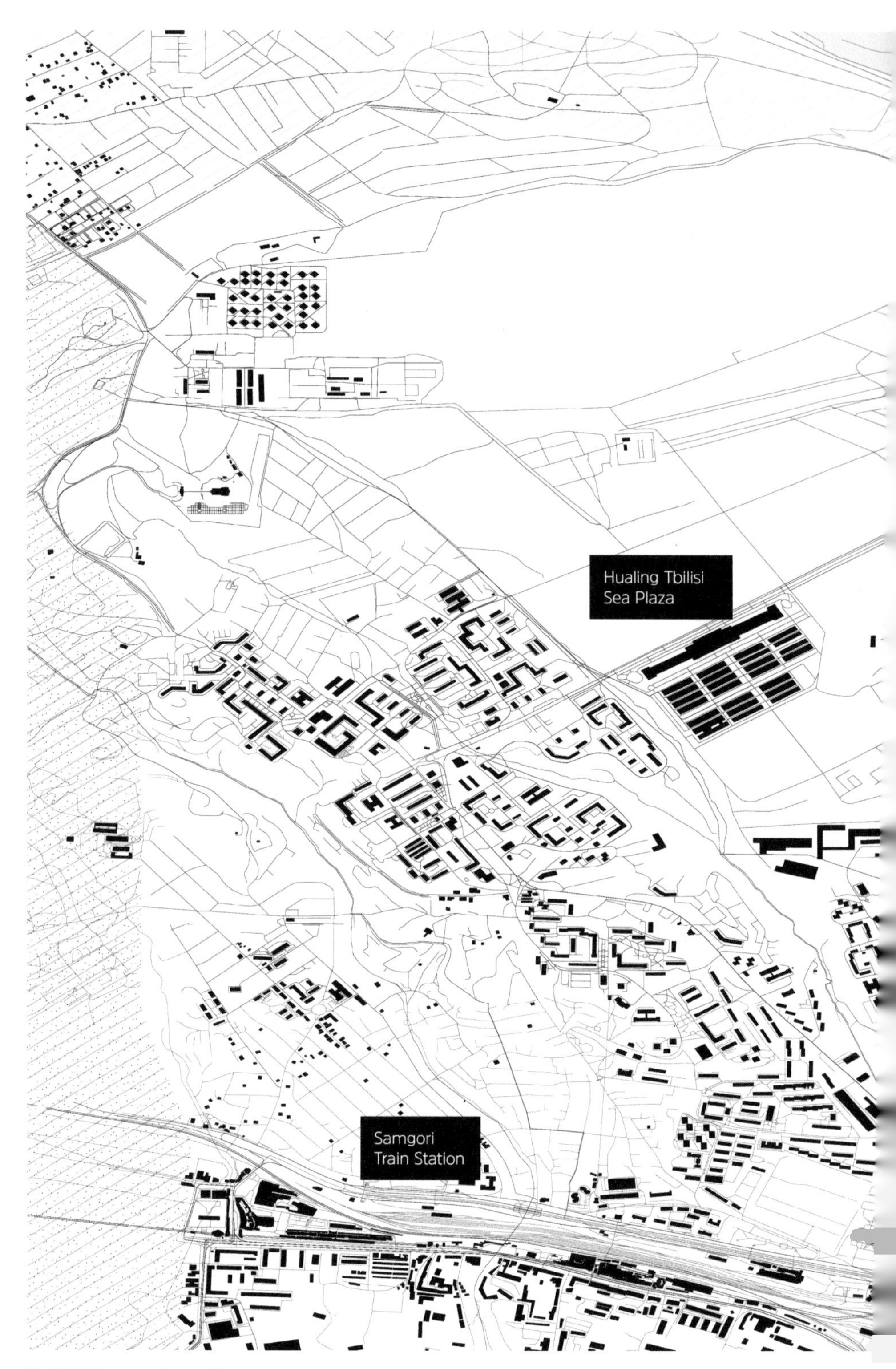

Territorial map on two scales.

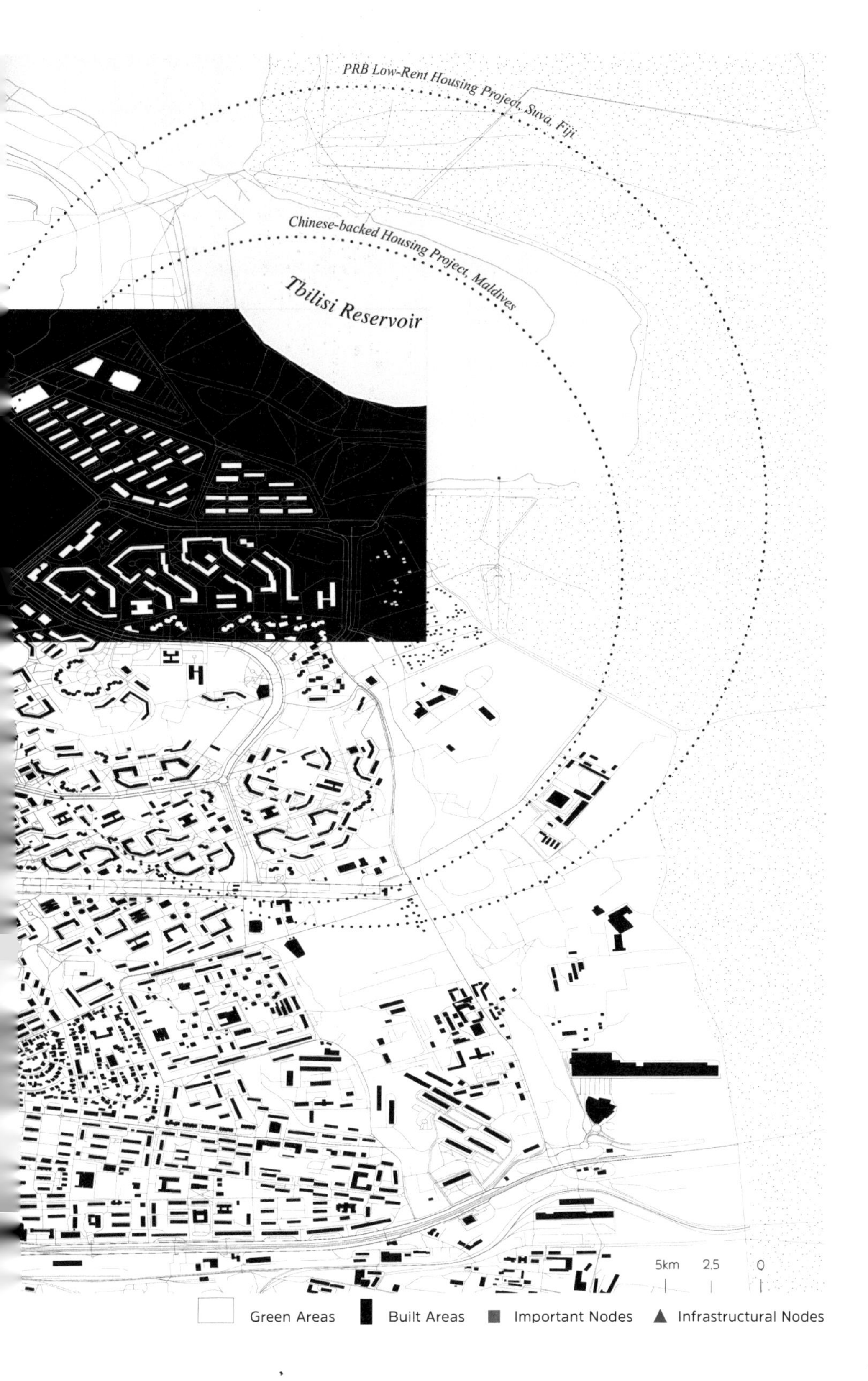
PRB Low-Rent Housing Project, Suva, Fiji
Chinese-backed Housing Project, Maldives
Tbilisi Reservoir
5km
2.5
0
Green Areas
Built Areas
Important Nodes
Infrastructural Nodes

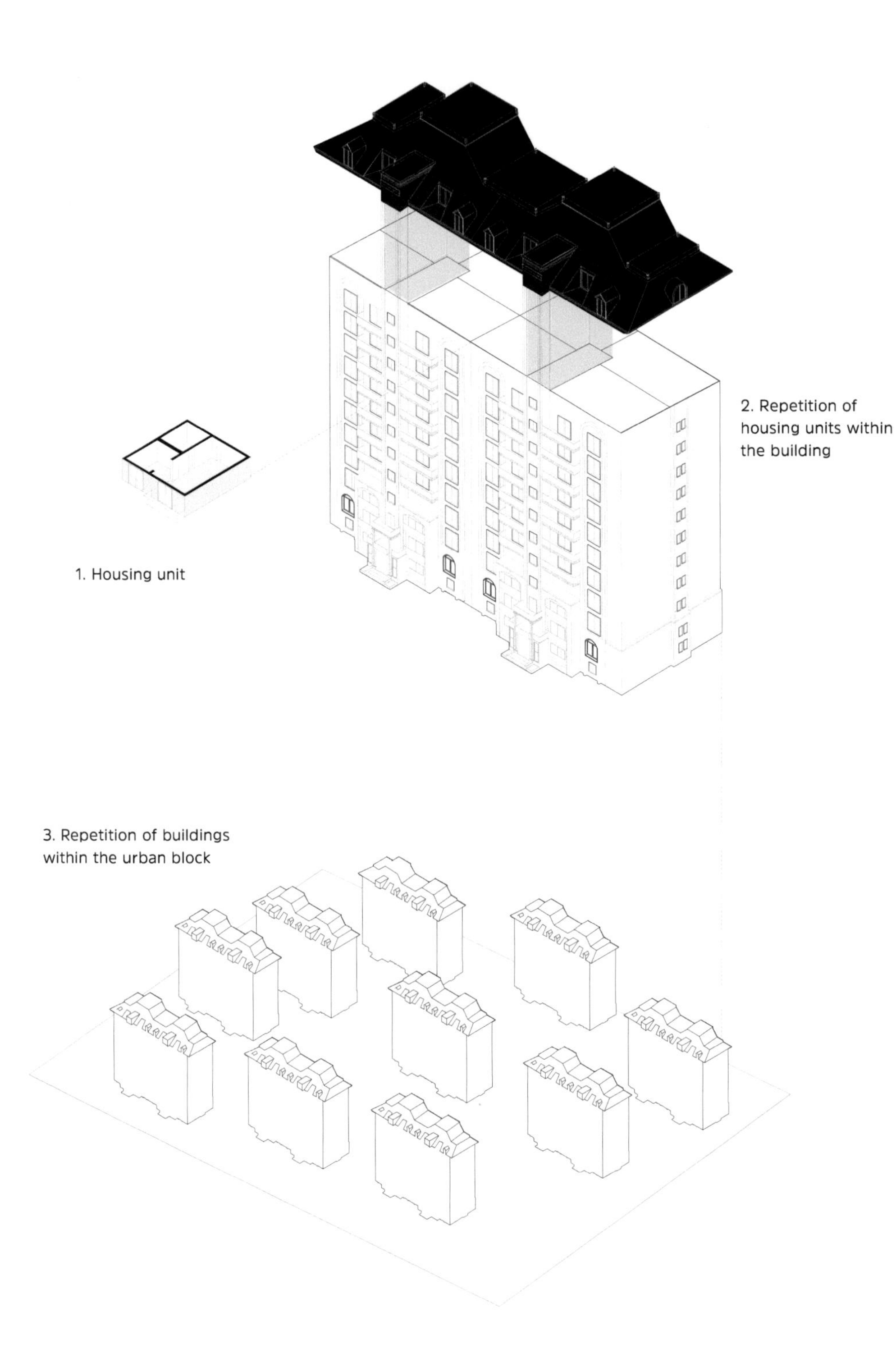

Morphological diagram showing the repetition of architectural elements at different scales.

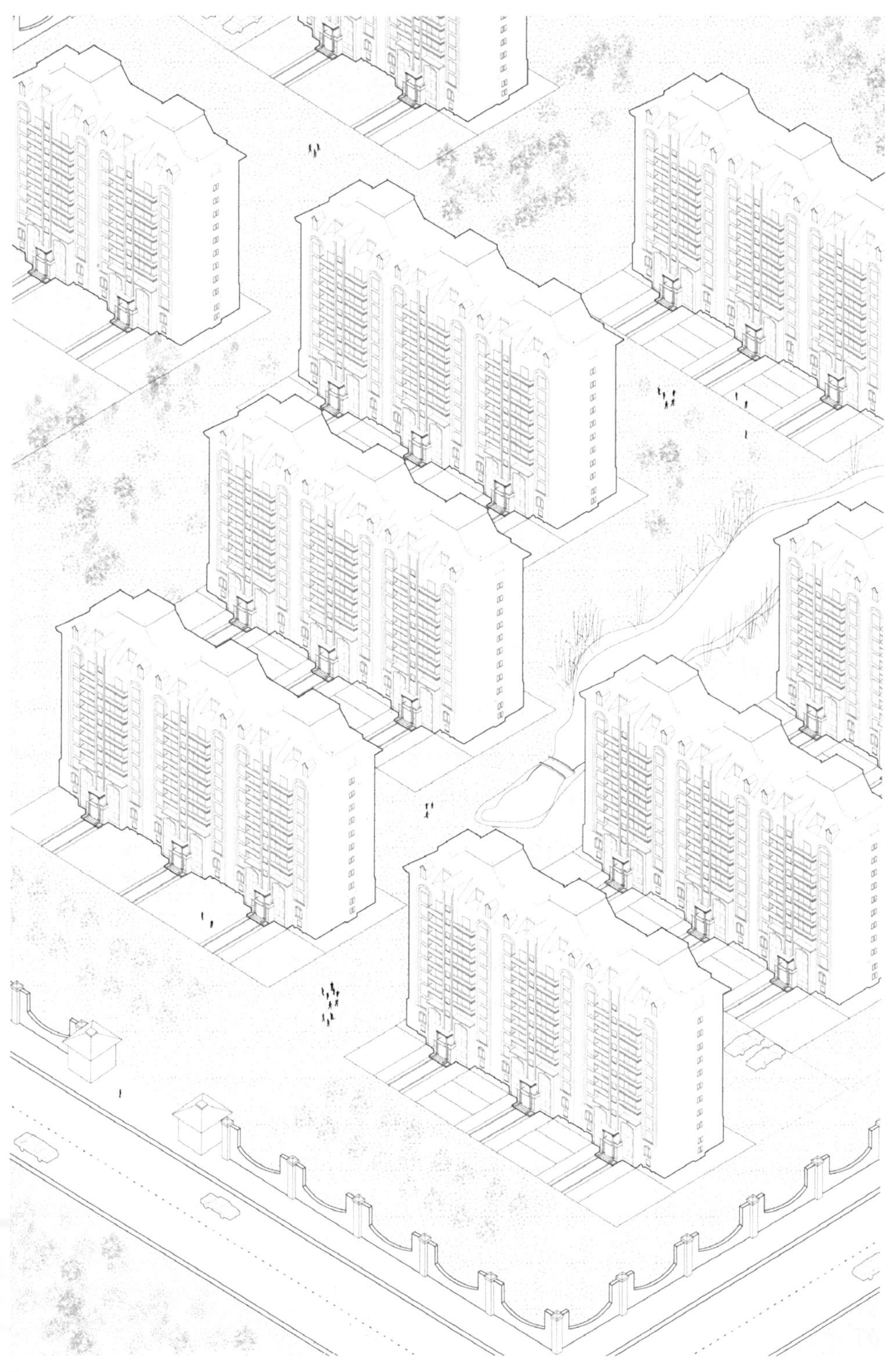

Axonometric view.

PRB Low-Rent Housing Project

Location	Suva, Fiji
Floor Area	10,240 square meters
Date	2018
Architect	China Railway First Group Co.,Ltd.
Client	Public Rental Board

The roof of the buildings, as well as the windows, are equipped with overhangs for rain protection.

Suva, the capital city of the Fiji Islands, hosts one of the biggest affordable housing developments in the Pacific Islands, supported and developed by the Public Rental Board (PRB), a local institution involved in the development and management of affordable housing. The PRB low-rent housing project of Raiwai flats in Suva is located in one of the most populated cities of the Fiji Islands; it allocates 210 apartment units to households with low incomes as part of a broader initiative to enhance the welfare of inhabitants and respond to the increasing demand for housing. The project – unveiled in August 2018 with financial support from China – saw the involvement of the China Railway First Group Co. Ltd. as the main project designer and contractor. A memorandum of understanding regarding collaboration under the Belt and Road Initiative (BRI) was signed in November 2016 by China and Fiji; at the time, Qian Bo, the Chinese ambassador to Fiji, highlighted China's commitment to the construction of infrastructure in the islands, including public-housing facilities.

Construction took almost two years; its contractual value amounted to approximately 73,434 million Chinese renminbi (RMB), roughly 10,215 million US dollars. The project covers a total built area of 10,240 square meters. The architectural program includes thirteen residential structures, a unique multifunctional assembly space, interconnected outdoor pathways, designated vehicular parking lots, and an integrated sports ground. The area, surrounded by luscious greenery and accessible from one of the main local roads, hosts an urban development that looks like a small village. Three- to four-story linear building blocks facing in different directions are positioned so as to generate dedicated outdoor spaces.

In order to make housing more affordable for future inhabitants, the project includes small apartments ranging from 50 to 70 square meters in which the size of each room is reduced to the minimum standard. The apartments are equipped with minimum facilities and decorated with standard Chinese elements and materials such as thin transparent sliding doors and polished faux marble tiles on the floor. Overall, the project reveals a strong functional image and approach to architecture so as to lower minimum construction time and post-occupancy management costs. In fact, the housing blocks are made of precast concrete elements covered by a simple plaster layer. All the windows and openings are the same size; they are completely prefabricated and assembled on site. An overhang pitched roof, clad with aluminum plates, protects the wall finishing from the strong rains that characterize the island's tropical climate. Even though the design of the buildings' façades is very ordinary, several specific, detailed solutions have been adopted to help conform the buildings to the regulatory and environmental context: a series of small projecting concrete elements overhang each window to protect them from the rain; all the apartments feature small stairs hung on the balconies for fireproof safety reasons, as required by local regulations; and a more resistant dark colored plaster has been used on the ground floor portion of the façade in order to tackle the floods that occur during the rainy season. Although these features appear to be minor details, they wield considerable influence in the context of such iterative spatial production.

Territorial map on two scales.

0
200 m
South Pacific University
Hualing Tbilisi Sea New City, Tbilisi, Georgia s
Kilamba City, Luanda, Angola
Laucala Bay
5km
2,5
0
Green Areas
Built Areas
Important Nodes
Infrastructural Nodes

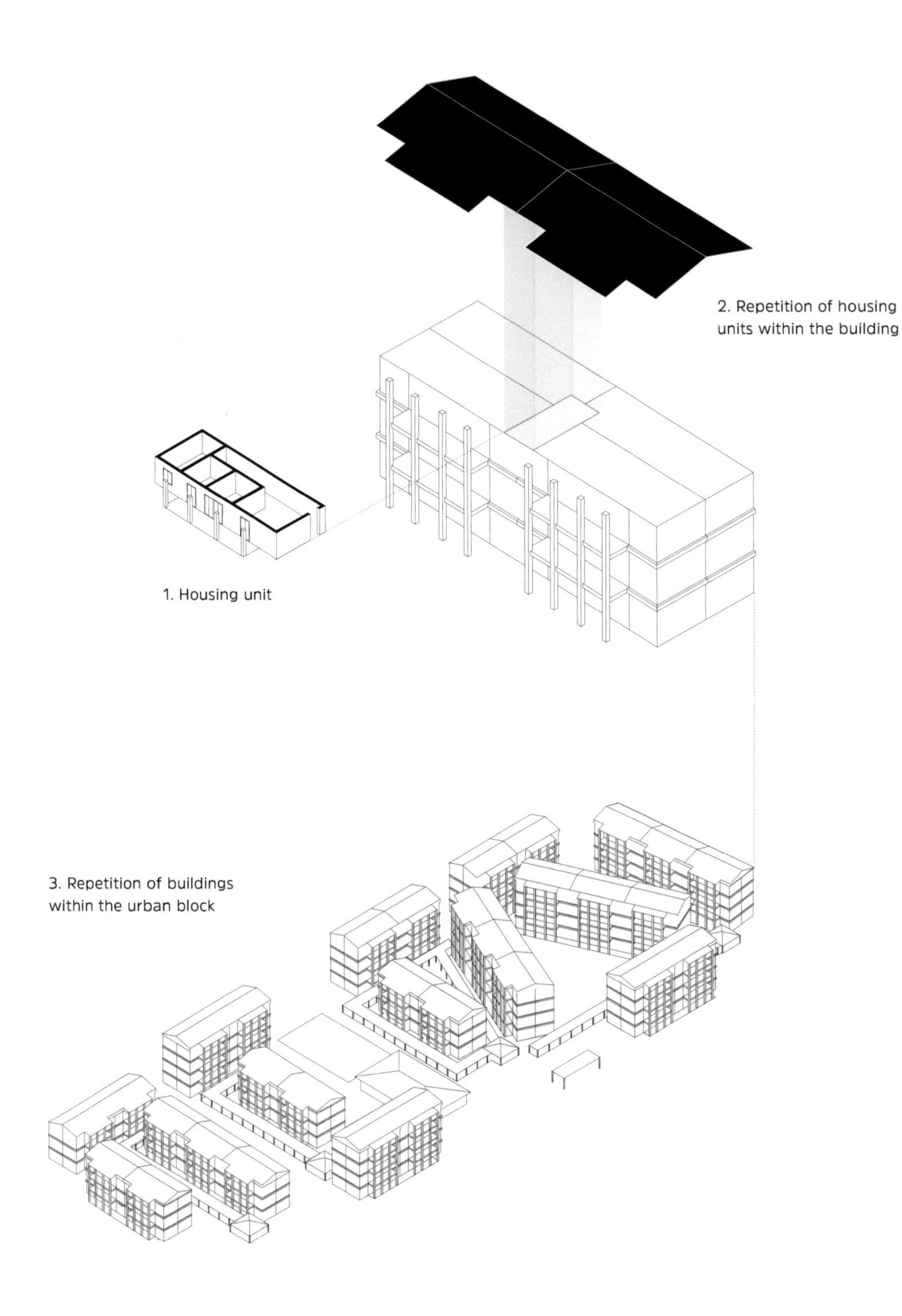

Morphological diagram showing the repetition of architectural elements at different scales.

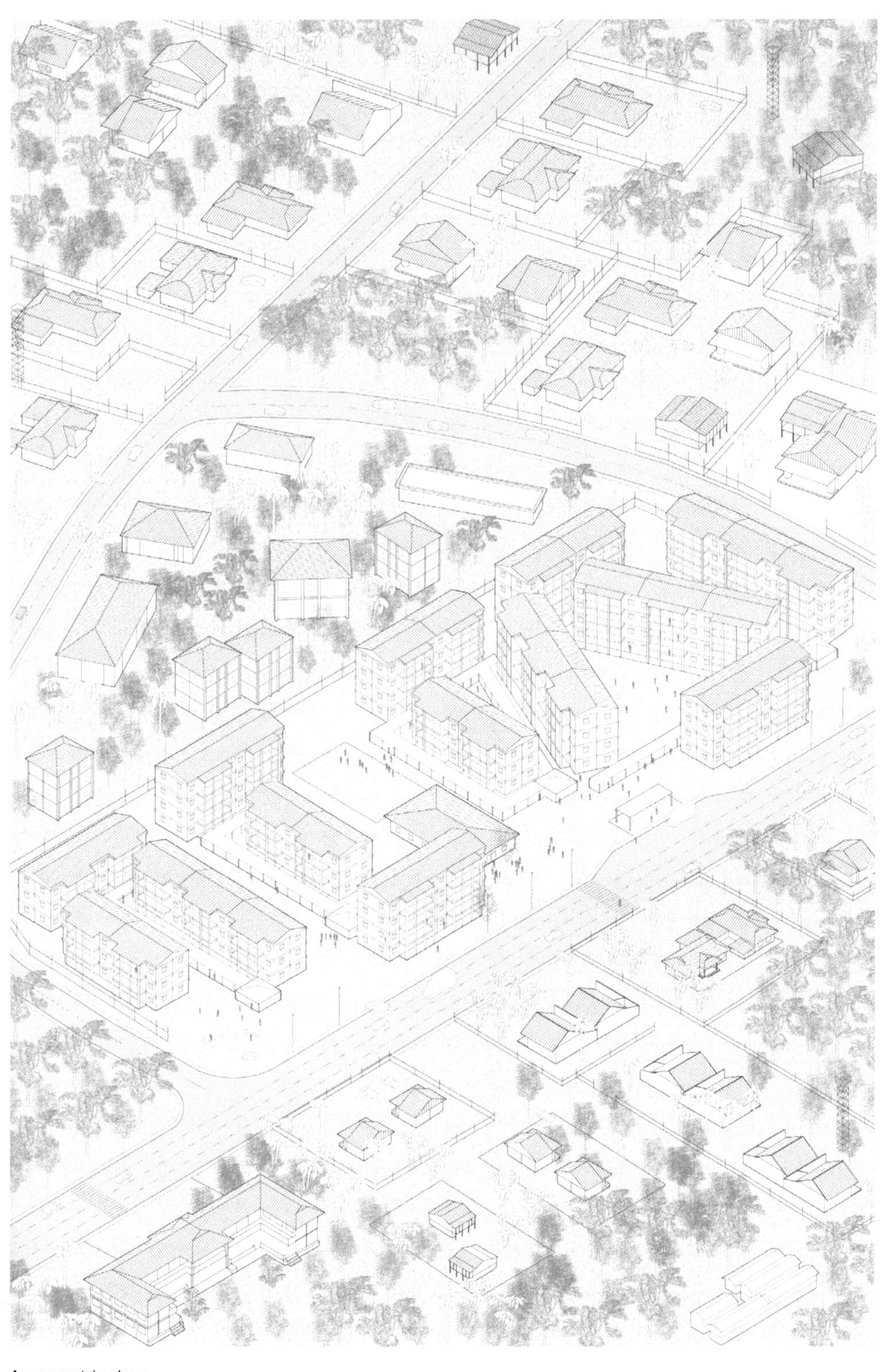

Axonometric view.

Kilamba Kiaxi Housing Estate

Location Luanda, Angola
Area 30.5 square kilometers
Date 2014

Buildings of various heights and forms are characterized by different colors in their façades.

Kilamba New City, also known as Kilamba Kiaxi or Nova Cidade de Kilamba, is a residential development located approximately 30 kilometers south of Luanda, the capital of Angola. The first phase of the project was completed in July 2012 with financial support from China. This flagship project, sponsored by the former Angolan President José Eduardo dos Santos (in power from 1979 to 2017), was touted as a solution to the housing deficit for an emerging middle class; it also included several social-housing units. The state-owned China International Trust Investment Corporation (CITIC) contributed to the development, which also received financial backing from the Industrial and Commercial Bank of China (an investment of 3.5 billion US dollars). The loan was to be repaid through Angolan oil sales.

Kilamba Kiaxi, with its 750 buildings and roughly 25,000 housing units, was intended to house around 500,000 residents; it is one of the biggest housing projects built by a Chinese firm on foreign soil, acting as a blueprint for new urban areas in Africa. The development is based on the Chinese new town planning model, characterized by precise boundaries, rigid infrastructure systems, and repetitive architectural elements. The project was divided into three phases that were to include not only housing estates, but also the establishment of essential services such as schools and kindergartens, as well as water, electricity, and sewerage facilities. Although several improvements have been made to the first ring road around Luanda, car dependency and lack of public transport continues to be a problem. While the city was initially described as a "ghost town" in the desert because of its substantial vacancy, as of 2023 it is almost fully occupied. Generally connected to water and electricity, the new development represents an attractive place for an emerging middle class (while 42% of Angolans still live without access to drinking water and 60% without sanitation). As of 2019, however, roughly half of the 20,000 home buyers in Kilamba were in arrears with their installment payments.

The general masterplan is based on a regular grid; dense residential plots are interspersed with open spaces to be used as public spaces by the inhabitants. The exterior colors of the 750 mid-rise and high-rise buildings, e.g., green, blue, and yellow, make them stand out in these plots. Three main types of residential buildings (from five to thirteen stories) contain three different apartment typologies, varying from three to five rooms and ranging from 110 to 150 square meters. All the residential towers have a central core, serving as a pivot around which standardized apartment units with identical layouts are either mirrored or rotated. These simply designed floor plans are then extruded vertically along the core without any variations between the floors; this keeps production costs and construction time to a minimum. The central core, enclosing the building's vertical distribution, is designed with a permeable mesh made of prefabricated concrete elements, thus facilitating natural cross ventilation, so necessary in the arid and hot climate of Angola. Any visitor who explores Kilamba City will notice the repetitive housing made up of apartment units and building layouts that do not vary in form and size.

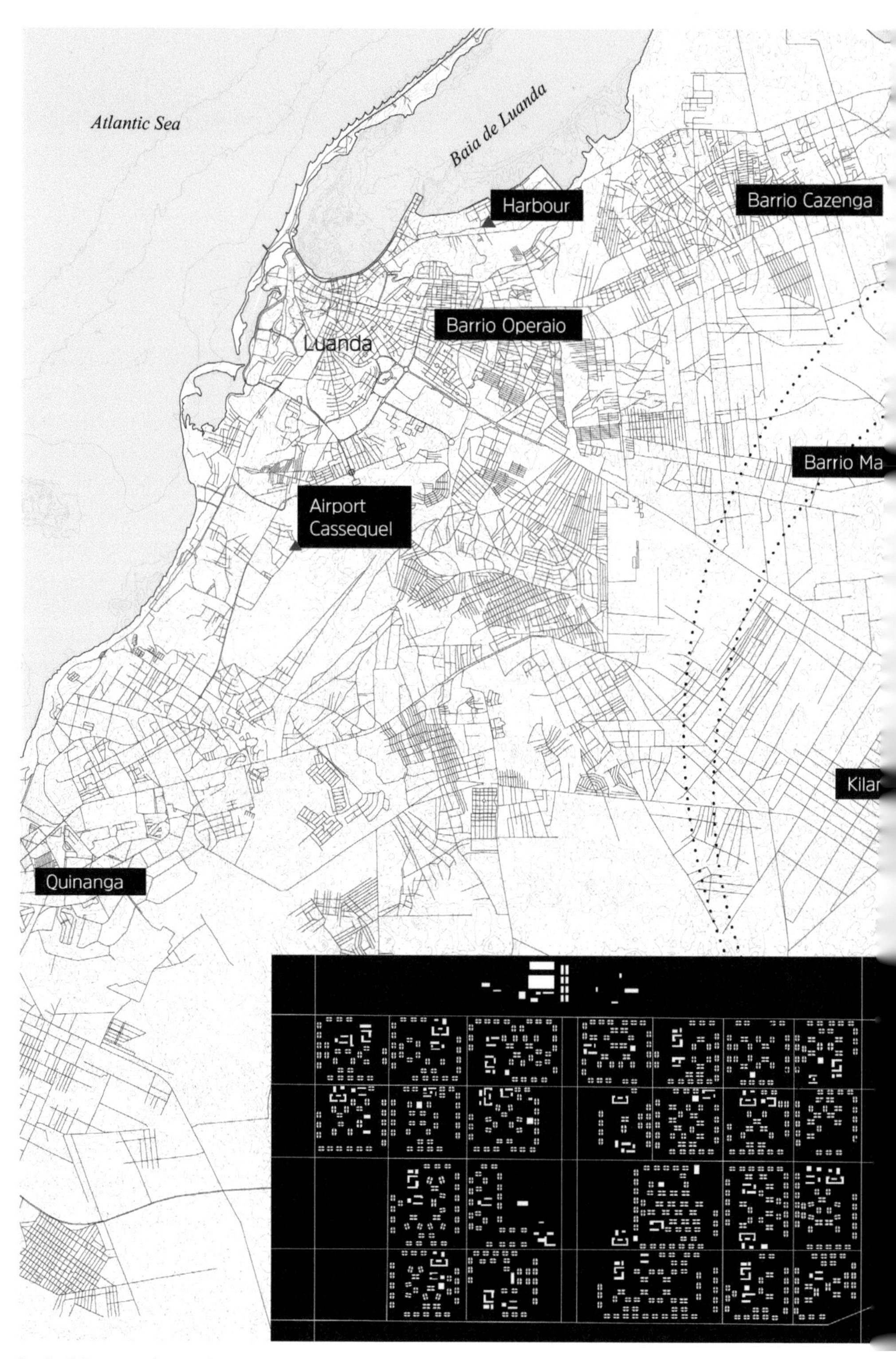

Territorial map on two scales.

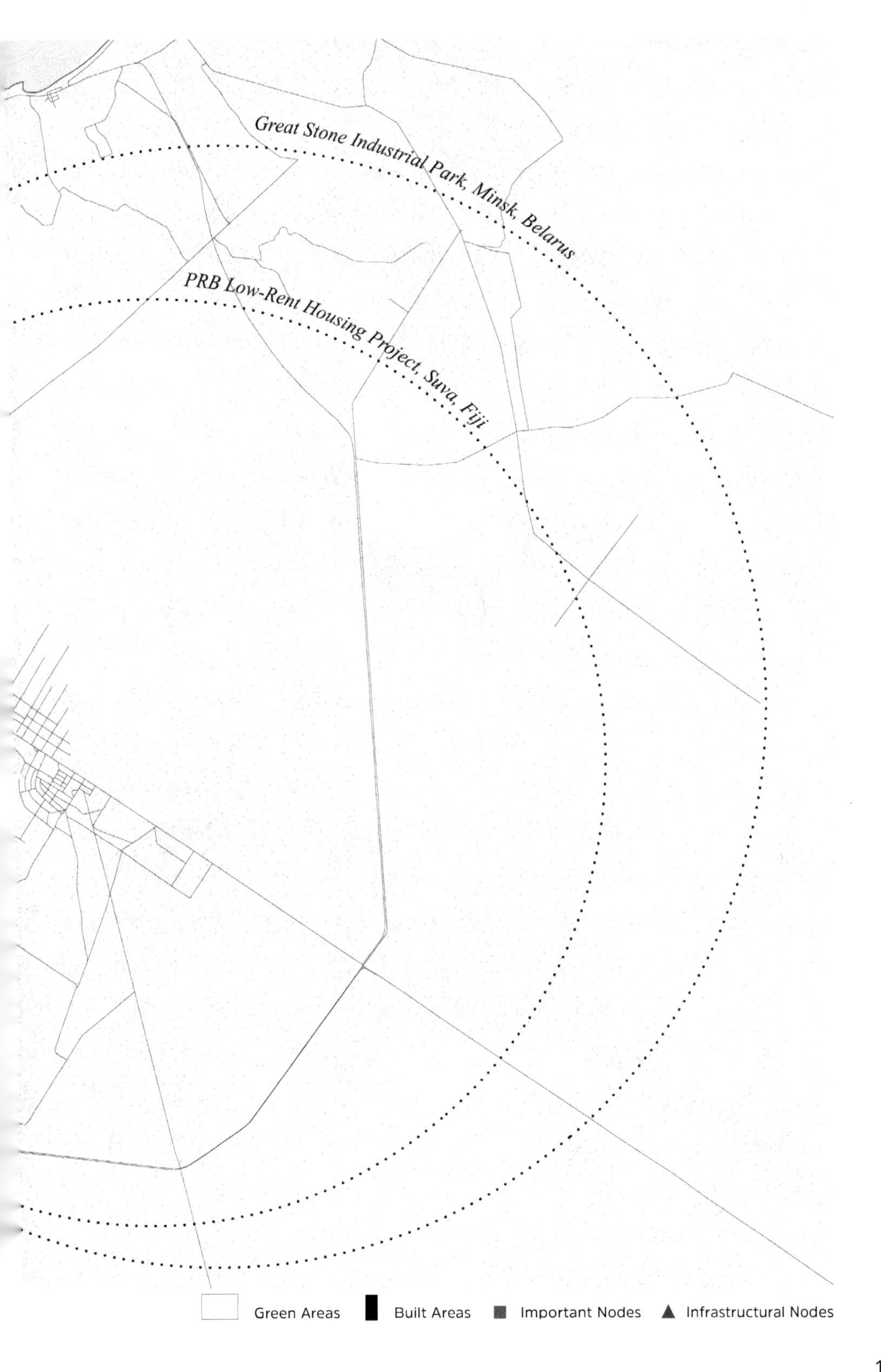
Great Stone Industrial Park, Minsk, Belarus
PRB Low-Rent Housing Project, Suva, Fiji
Green Areas
Built Areas
Important Nodes
Infrastructural Nodes

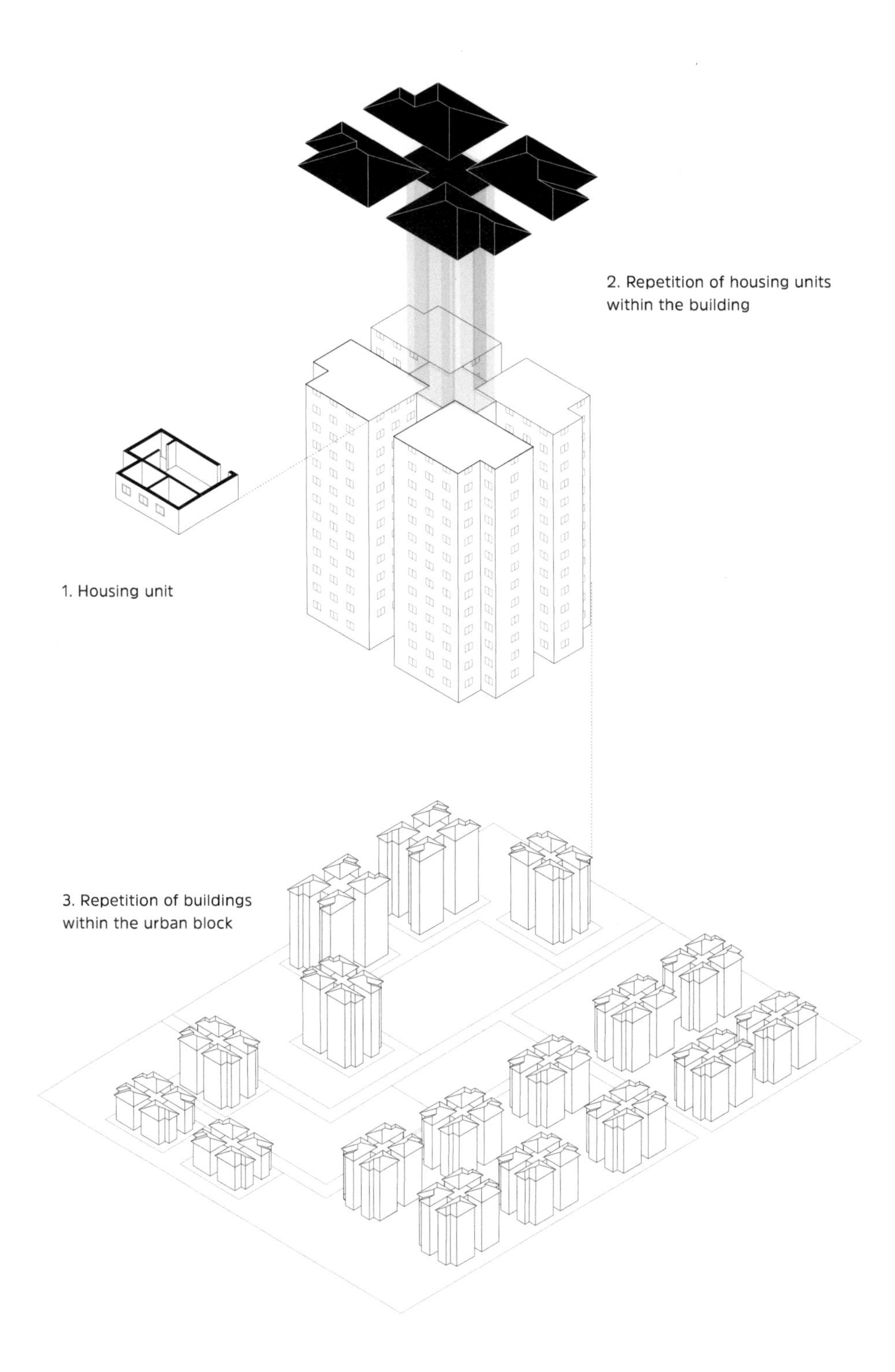

Morphological diagram showing the repetition of architectural elements at different scales.

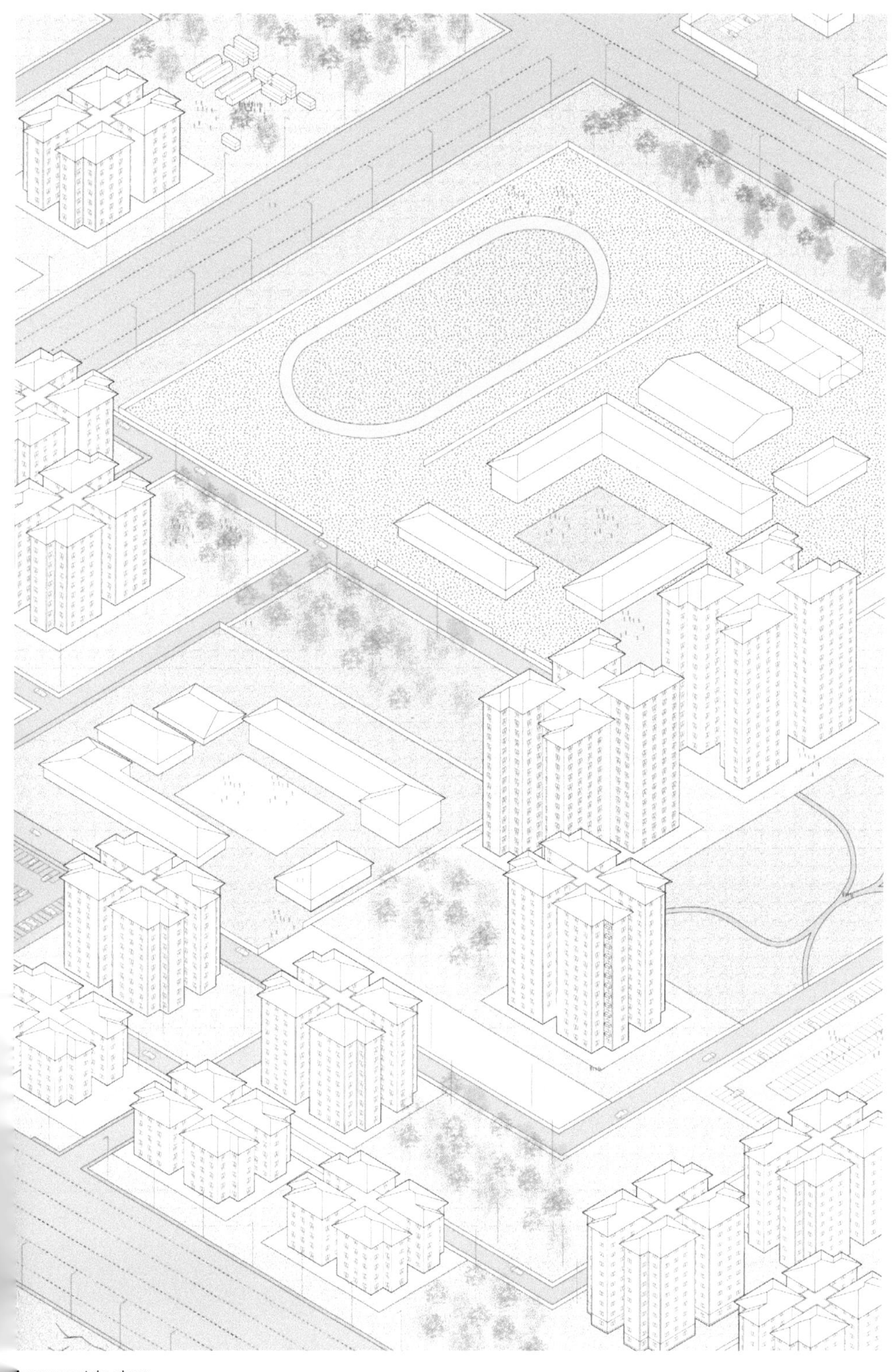

Axonometric view.

Super Gathering Places

Meeting in between Architecture

Xi'an Silk Road International Convention and Exhibition Center

Location	Xi'an, China
Floor Area	181,200 square meters
Date	2022
Architect	gmp Architects
Client	Xi'an Company Silk Road International Convention and Exhibition Center

Glazed curtain walls form the envelope of the complex creating an impression of weightlessness from the exterior.

The Silk Road International Conference Center, located in the north-east area of the city of Xi'an, reflects the importance of the Belt and Road Initiative regarding cultural and multi-purpose venues. The conference center is part of a larger complex, designed to host international events, trade fairs, and congresses. It is important to recall that Xi'an is the city where the old Silk Road began before it crossed the whole of inner China and ended in Europe. The conference center was designed by the German office gmp Architects after an international competition launched by the Xi'an Company Silk Road International Convention and Exhibition Center.

Gmp Architects designed a building that echoes a previous era of 20th-century China, but with a contemporary twist: a period of "big roofs" and a search for a national style comprising traditional elements stripped of their history. References to traditional Chinese architecture can indeed be seen in the symmetrical shape of the roof, in the proportioned façades with a mainly horizontal development, and finally in the thin 180 columns that surround the building and imbue it with natural light. The result is a technologically advanced, contemporary building with references to the local architecture. Symbolically speaking, the building is a "temple" where people from all around the world come and meet. As in traditional Chinese buildings, the suggestive exterior columns surrounding the big volume all along its perimeter are both ornamental and functional. Here, the columns, hanging from the cantilevered roof beams, allow the lower half-moon-shaped arch of the façade to be suspended, thus clearing the entrance from the structure and generating a sense of weightlessness that characterizes the building's appearance. The intricate structural system of the complex, made up of a grid of steel beams spanning the cores, provides a large flexible space, free from any obstructing load-bearing element, thus making it ideal as a venue to host large-scale events. This flexible space allows the program to be organized on three floors of the building where the ballroom, the multifunctional hall, and the conference space are respectively located. Each floor is accessed by a U-shaped corridor, functioning both as a foyer and a space for temporary exhibitions.

Lightness, openness, and harmony characterize the interior space where light floods in from the glazed curtain walls on all sides of the building. This façade - the largest steel mullion-supported curtain wall in China - provides the building with a sense of transparency and lightness. Its strong structural language weaves past and present together in a fabric of references, generating cultural bridges and symbolic messages, and thus enhancing the culture around the New Silk Road.

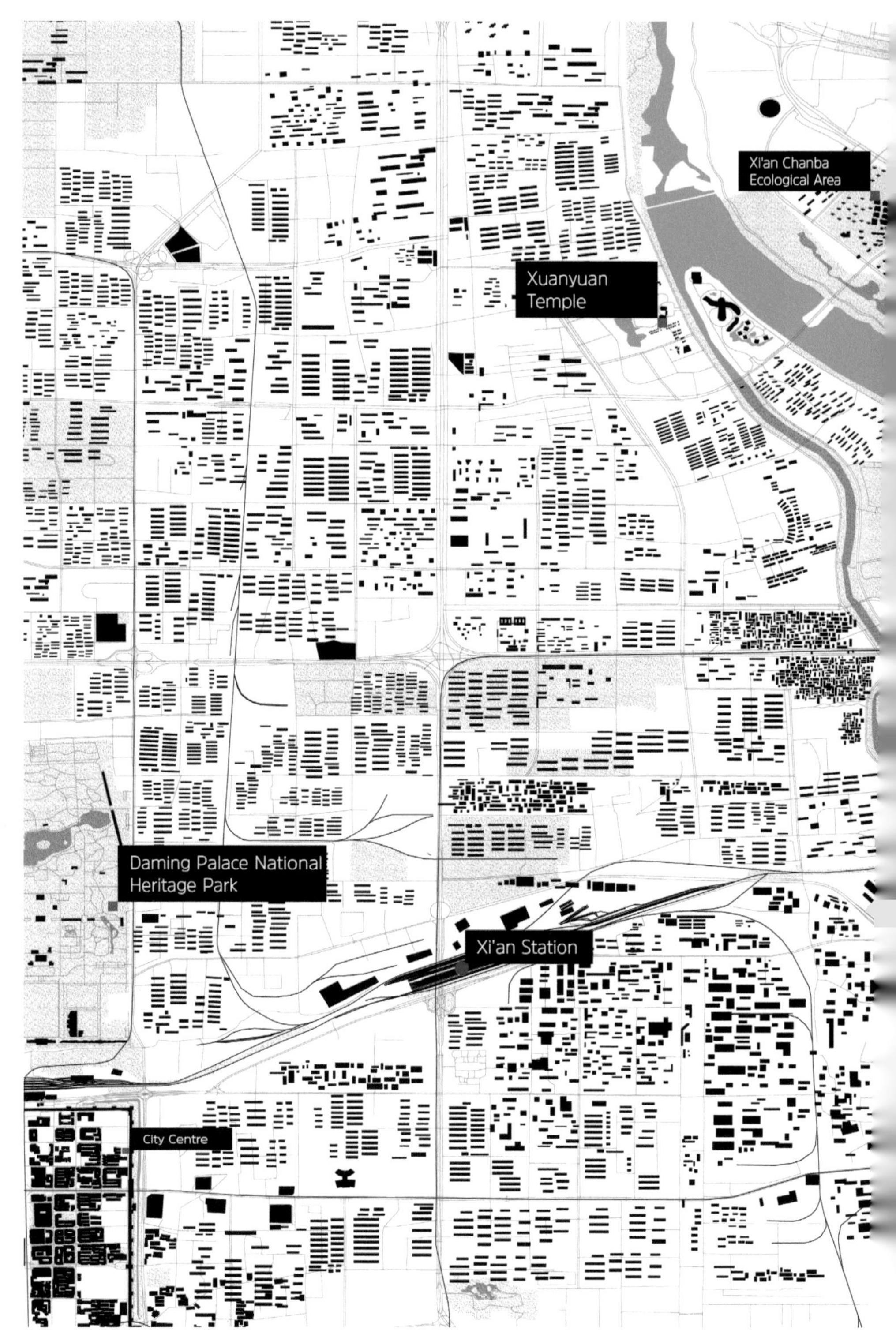

Territorial map on two scales.

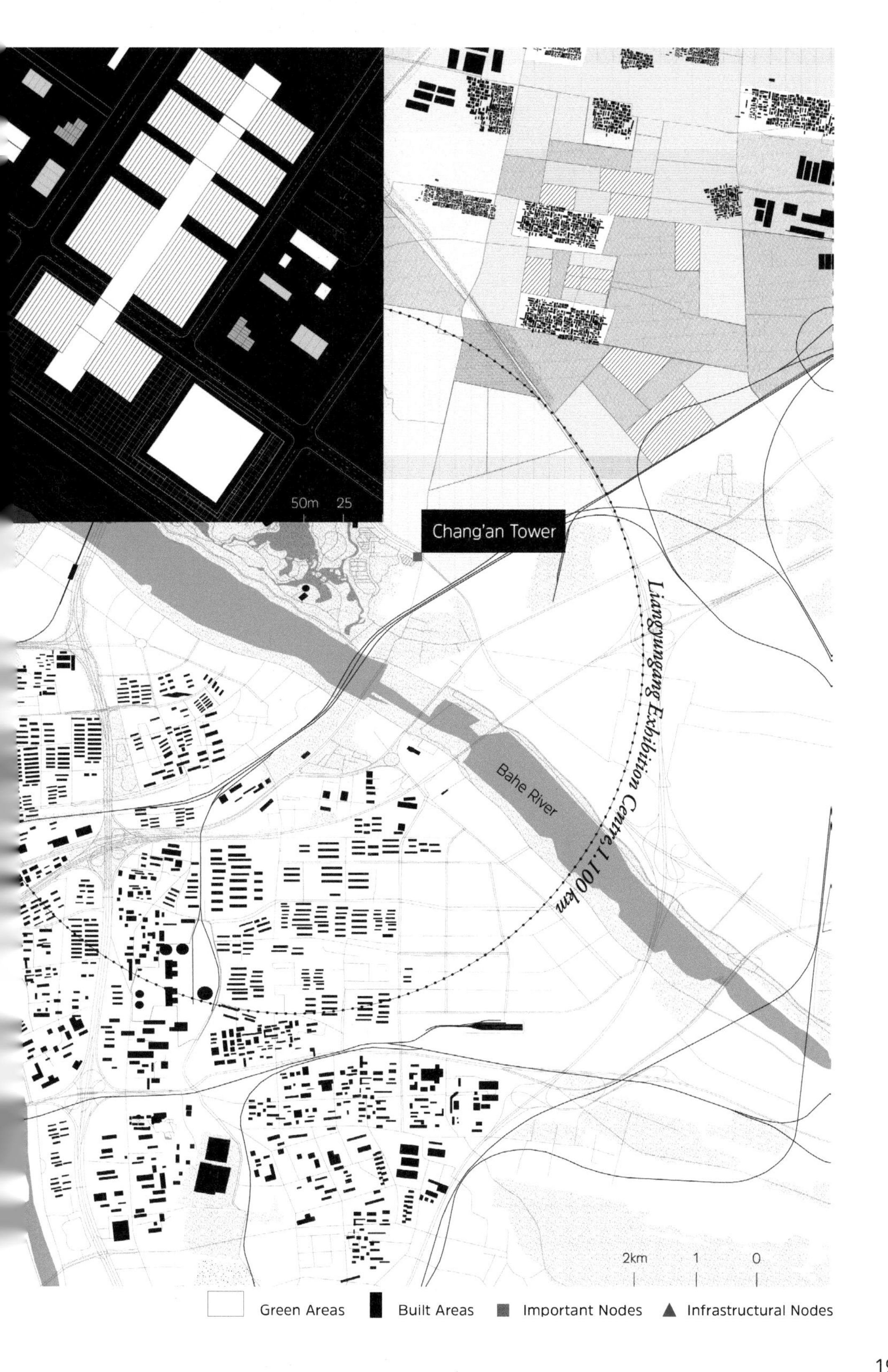
50m
25
Chang'an Tower
Lianyungang Exhibition Centre, 1,100 km
Bahe River
2km
1
0
Green Areas
Built Areas
Important Nodes
Infrastructural Nodes

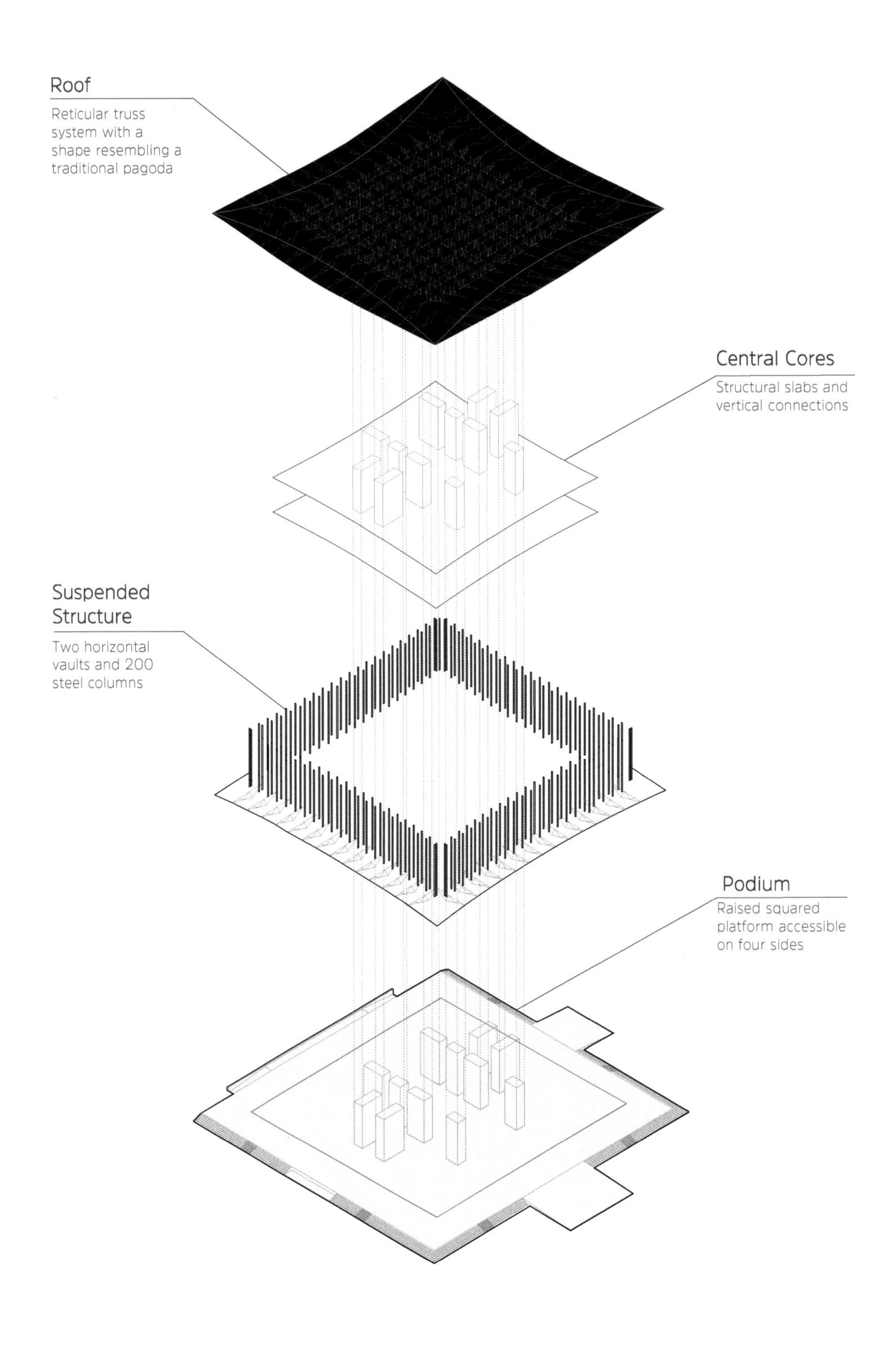

Relation diagram showing the main elements composing the tectonic of the building.

Axonometric view.

Lianyungang Industrial Exhibition Center

Location Lianyungang, China
Floor Area 64,000 square meters
Date 2017
Architect gmp Architects
Client Lianyungang Rui Hao Investment and Development Co., Ltd.

Vertical incisions in the façade resemble the shape of a barcode.

The Lianyungang Industrial Exhibition Center is located close to the Yellow Sea in the northern part of the city of Lianyungang, situated in the eastern area of the new Eurasian Land Corridor established under the Belt and Road Initiative. Designed by von Gerkan, Marg and Partners (gmp Architects), the project reflects the city's trading and commercial tradition.

The exhibition center hosts a simple program consisting of four large exhibition areas for trade shows, concerts, and other events, organized on two stories similar in form and dimension. The four exhibition halls are positioned at the four corners of the complex, defining the total volume of the building with its 200 × 320 meter footprint area. The clear, functional circulation space between the halls leads to four entrances: from the north and south for delivery access, and toward the east and west for pedestrians. All the entrances to the building are recognizable from its exterior due to tiered incisions in the building's structure – a reference to the Dougong principle of traditional Chinese roof construction. A big central staircase in the core of the building provides access to the center's upper level; gmp Architects have used the staircase as a functional structure in which to hide the bathrooms and service spaces. The monumental proportions of the vertical distribution result in a visual and spatial continuum between the two floors of the building. Cascading water features positioned between the staircases characterize the space and create an atmosphere reminiscent of oriental gardens. Two meeting rooms are located in front of the upper entrances to the main halls.

From a structural and tectonic point of view, the distribution space and the four halls are autonomous elements. The load-bearing structure uses mixed concrete and steel truss beams to create the wide span of the building. The massive exterior façade is broken up by a series of skylights allowing the sunlight to enter from the top: the result is an expansive well-lit space lending itself to all kinds of events. The massive, solid façade with its natural stone cladding panels features a non-continuous vertically staggered incision reminiscent of a barcode – a symbolic reference to the function of the building as a place for trade and commerce. At night, vertical LED lighting illuminates the "barcode," bestowing an unmistakable identity on the building within the urban context.

Territorial map on two scales.

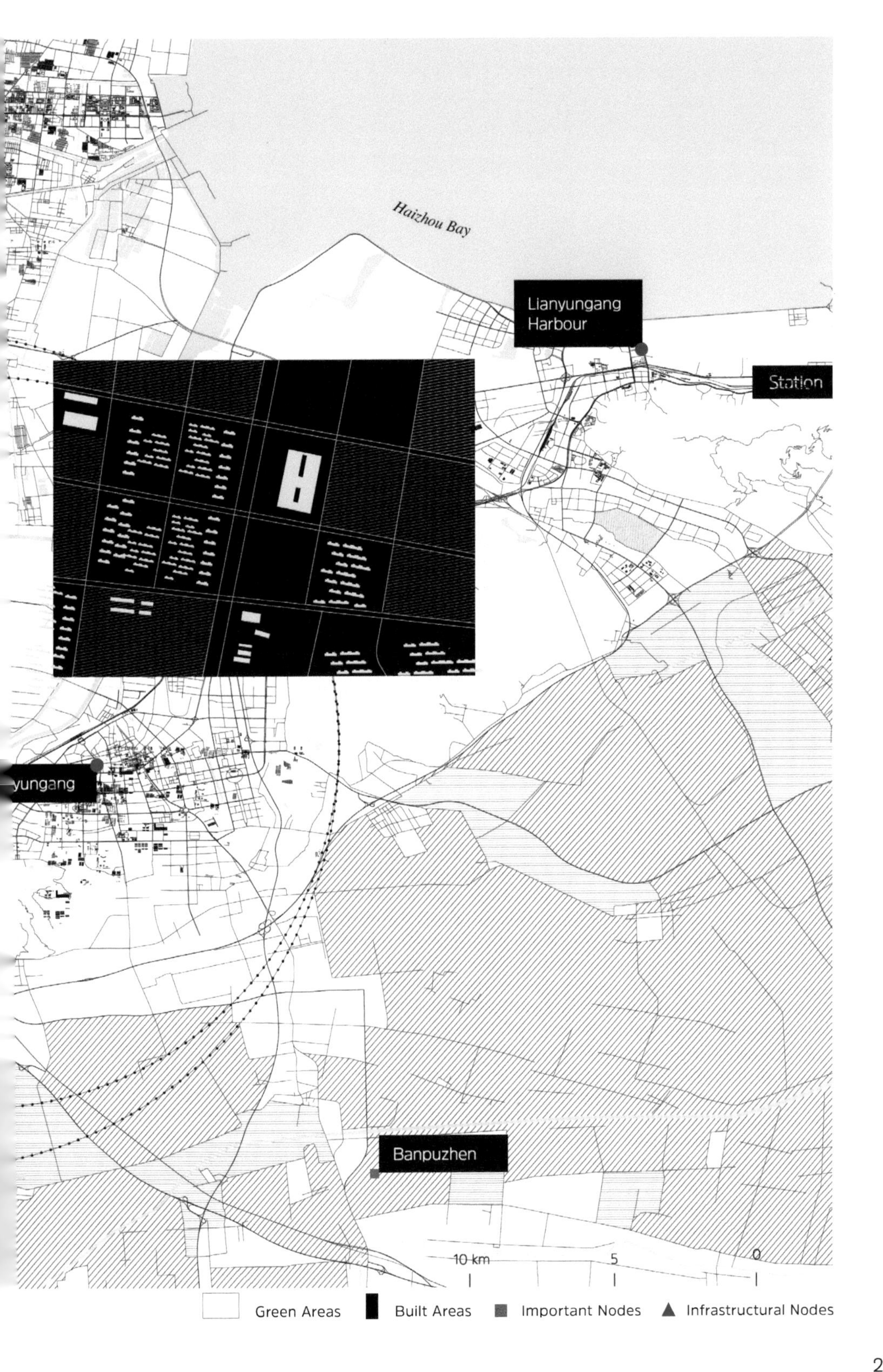

Haizhou Bay
Lianyungang Harbour
Station
yungang
Banpuzhen
10 km
5
0
Green Areas
Built Areas
Important Nodes
Infrastructural Nodes

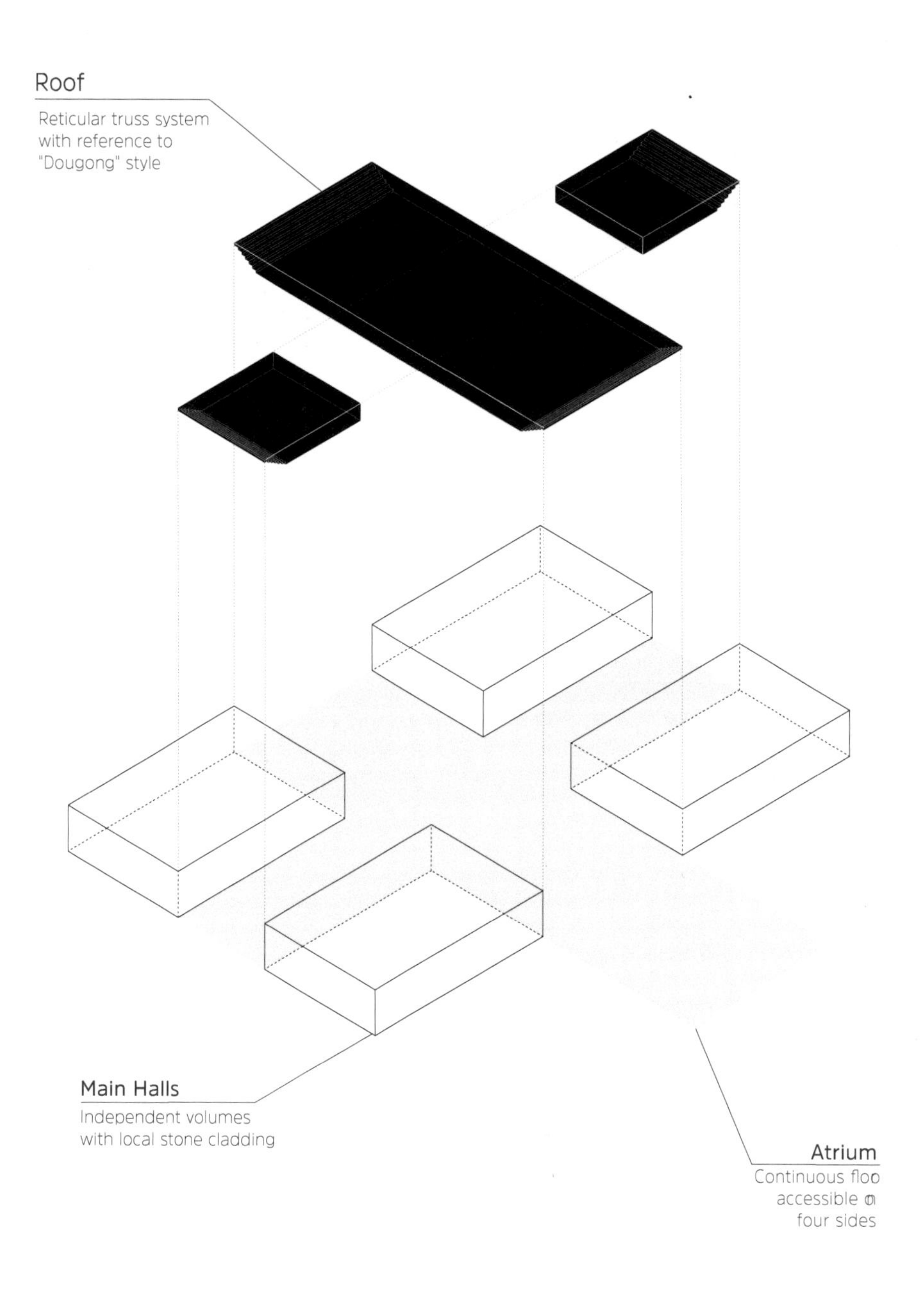

Relation diagram showing the main elements composing the tectonic of the building.

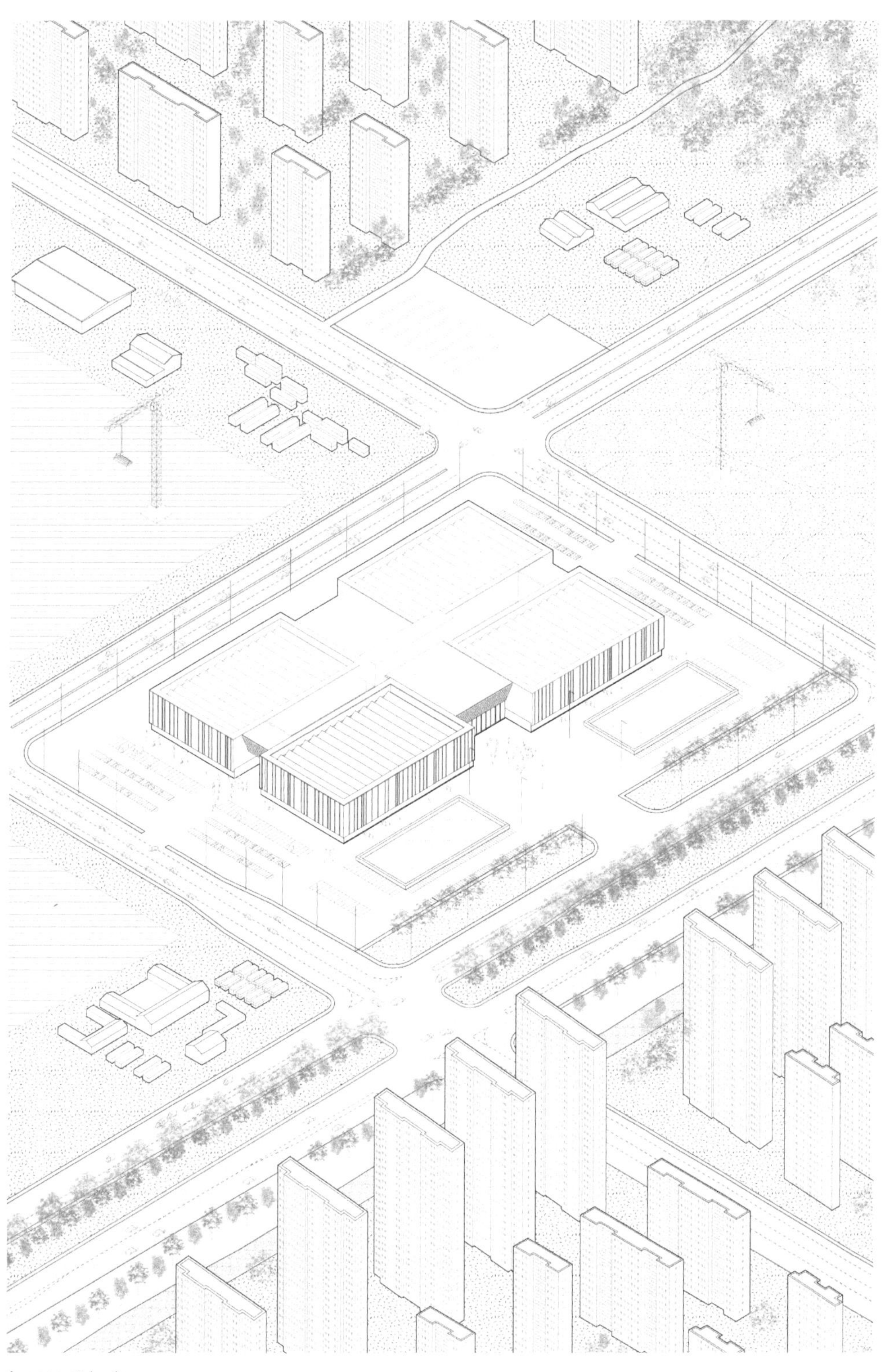

Axonometric view.

Langfang Silk Road International Cultural Exchange Center

Location	Langfang, China
Floor Area	34,000 square meters
Date	2021
Architect	Chidori Yoshinori, WAY Studio
Client	ENN Group

The envelope made of curvilinear metal louvers generates a fluid and dynamic architectural experience.

The Langfang Silk Road International Cultural Exchange Center is located a few kilometers south of Beijing; it is one of the new northern Chinese cultural hubs implemented as part of the "Dreams of Langfang" initiative, a government program designed to assimilate, foster, and promote cultural resources in relation to environment-friendly initiatives. As part of this megaproject, the Silk Road International Cultural Exchange Center, with its maximum capacity of 1,800 visitors, can host numerous cultural activities, such as art exhibitions, concerts, and lectures.

Upon entering the sprawling plaza in the front of the building, the big center appears as an amorphous mass of light rather than a real building. However, once inside, the visitor realizes that the external appearance of the building - one big volume made up of a unique sparkling envelope of curvilinear metal louvers - does not reflect its more fragmented and nuanced program. It includes three different main areas: the center with its mixed-use cultural complex hosting an opera house, a theater, and a music hall; the east side with its exhibition hall; the west wing with, in the middle, an art museum recently renovated by the Chinese architecture firm WAY Studio. Commercial inserts are also present throughout. The ground floor functions as a big platform that connects the different areas, providing them with a unique hall and diversified circulation routes, thus allowing each area to work either individually or together as an entire complex.

The big atrium offers an impressive overview of the building's tectonics: the sun shading system made of adjustable, continuous horizontal blades that surround the whole volume, allow the natural light to filter from the skylights above. Throughout the day, light and shadow shift across the blank white surfaces in a sort of perpetually evolving performance. Fluidity and motion are appropriate terms to describe the architectural experience of the building, not only regarding its atmosphere, but also spatially. A continuous wavy balcony system, connecting visitors to galleries at both ends, emphasizes the relationship between seeing and being seen. Articulated platforms on all sides offer views back into the atrium center, similar to theater boxes overlooking a central stage. In this sense, the whole experience is a continuous fabric of moving views and images. While the spatial distribution is organized and clearly legible, the circulation flows allow for a varied use. The interior paths steer the visitor to observe the space from different viewpoints, offering a non-static experience. As a result, the tripartite planimetric layout is no longer noticeable, leaving room for the fluidity of the visitors' movements.

The unusual architectural appearance of the Langfang Silk Road International Cultural Exchange Center creates a contrast with its program. The building's symmetrical and monumental exterior composition is juxtaposed against the curved and spiraling interior.

Territorial map on two scales.

Pakistan China Friendship Centre, Islamabad, Pakistan
Lianyungang Exhibition Centre, Lianyungang, China
10 km
5
0
Green Areas
Built Areas
Important Nodes
Infrastructural Nodes

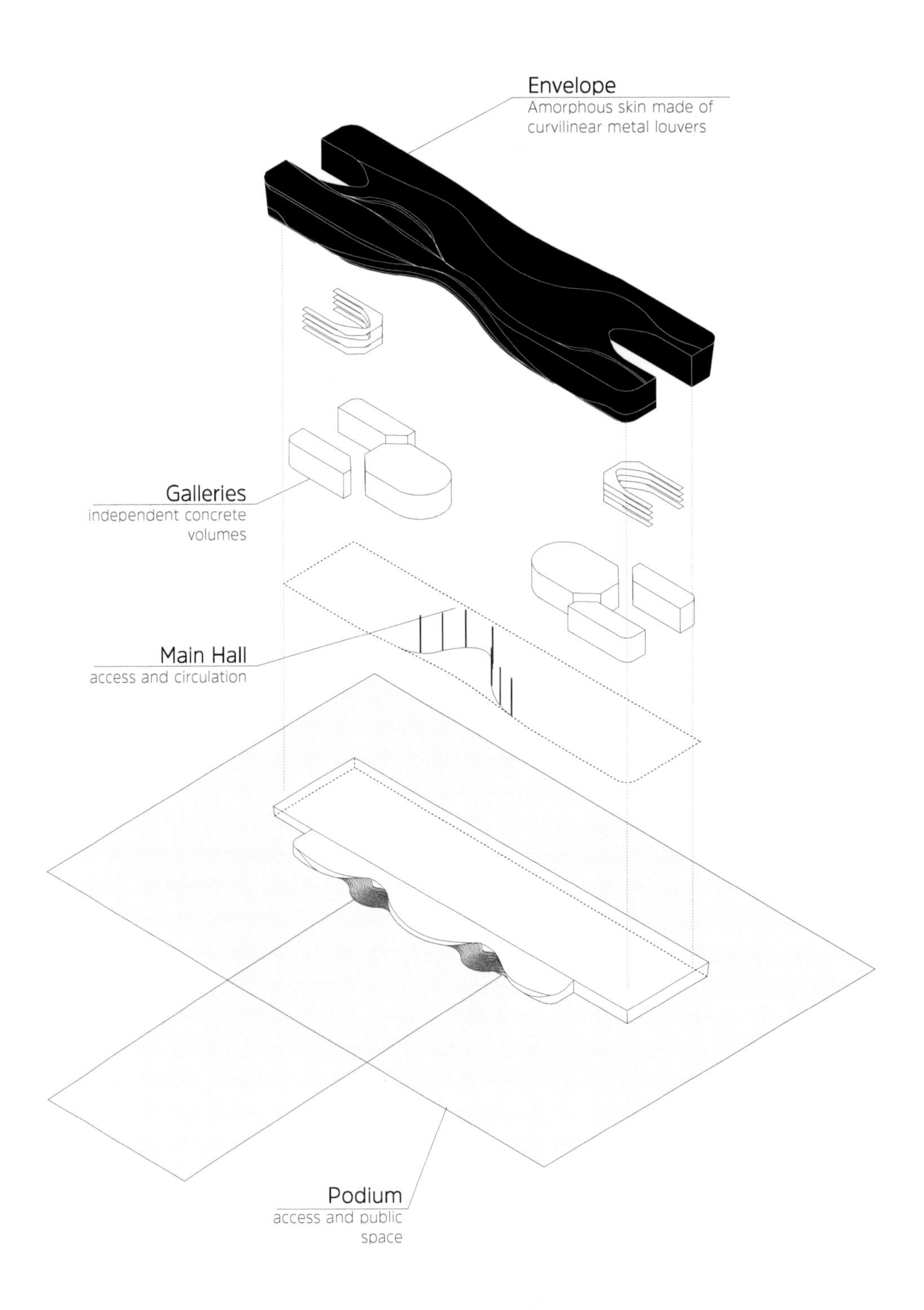

Relation diagram showing the main elements composing the tectonic of the building.

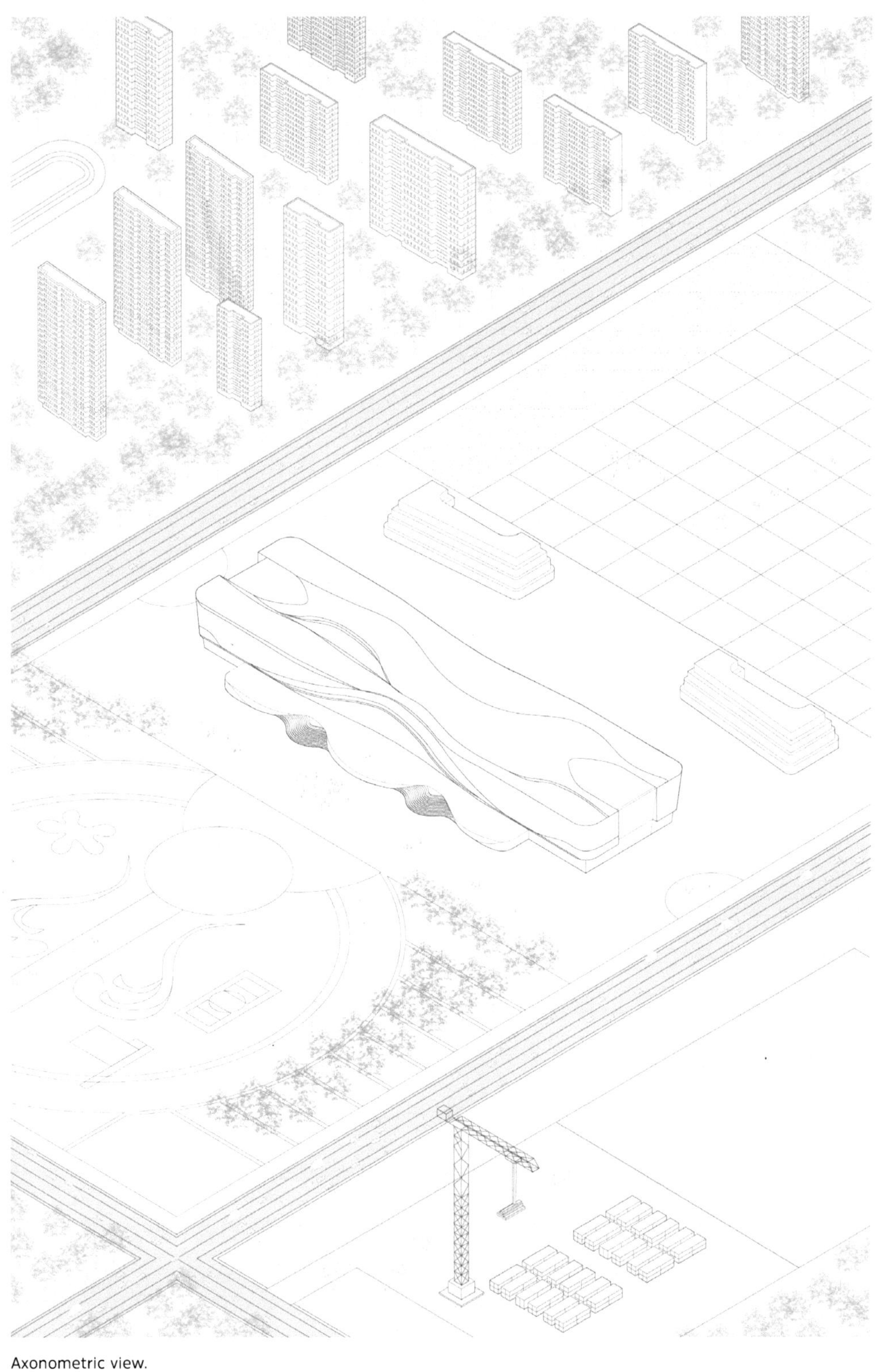

Axonometric view.

Bangladesh-China Friendship Exhibition Center

Location	Dhaka, Bangladesh
Floor Area	33,000 square meters
Date	2021
Architect	Beijing Institute of Architectural Design
Client	China State Construction Engineering Corporation (CSCEC)

The façade made of red terracotta tiles is protected by a suspended big roof.

Situated close to Bangladesh's capital of Dhaka, the Bangladesh-China Friendship Exhibition Center is a prime example of mutual cooperation under the Belt and Road Initiative. Inaugurated in 2021 by the Prime Minister of Bangladesh, the exhibition center is a permanent venue for different product-based fairs held throughout the year. The building is located in a newly developed area in the northeast part of Dhaka, covering 8 hectares of greenfield. The project, which enjoyed 48% financing by China, was built by the China State Construction Engineering Corporation (CSCEC) and designed by the Beijing Institute of Architectural Design.

The program of the exhibition center is quite complex and variegated: it has two main halls each with 400 exhibition booths, a 600-square-meter multifunctional hall, a dining room that can seat 500 people, offices, a prayer room, a staff dormitory, a children's activity area, and related functional auxiliary rooms. The programmatic and distributive structure of the center is quite obvious and straightforward; its symbolic dimension, however, is of great significance and was emphasized by the Chinese design team. Two architectural elements are joined in an expressive gesture: a main volume made of red ceramic panels and an aluminum gray roof, symbolizing the "ship" of friendship and trade between China and Bangladesh. As in other "Super Gathering Places," it is the roof in particular that plays an important aesthetic as well as functional role. In this case, the roof curves upward to cover the two stories of the center, creating a double-height space between the two main halls and allowing the plentiful natural light that comes through the outer curtain walls to fall into this space.

When viewed from the outside, the wavy roof structure has an exaggerated scale compared to the height of the building, thus turning it into a grandiose gesture. The building's monumentality is also enhanced by the enormous plaza in front, adorned with a rectangular pool and an array of flag poles organizing the open space. While the continuous roof protecting the spaces below is reminiscent of the curved terracotta structures of ancient Bangla temples, the main architectural composition may have been influenced by the narrative of the "big roof" that has emerged in recent Chinese architectural production. Bearing this in mind, it looks like a Chinese feature merging local traditional architectural elements with the political message of a "common destiny" within the BRI's rhetoric.

The Bangladesh-China Friendship Exhibition Center represents one of the most significant projects implemented within the BRI framework in terms of language, structure, and space. Since October 2021 Dhaka has enjoyed its new fair center which has become quite popular among local and foreign visitors.

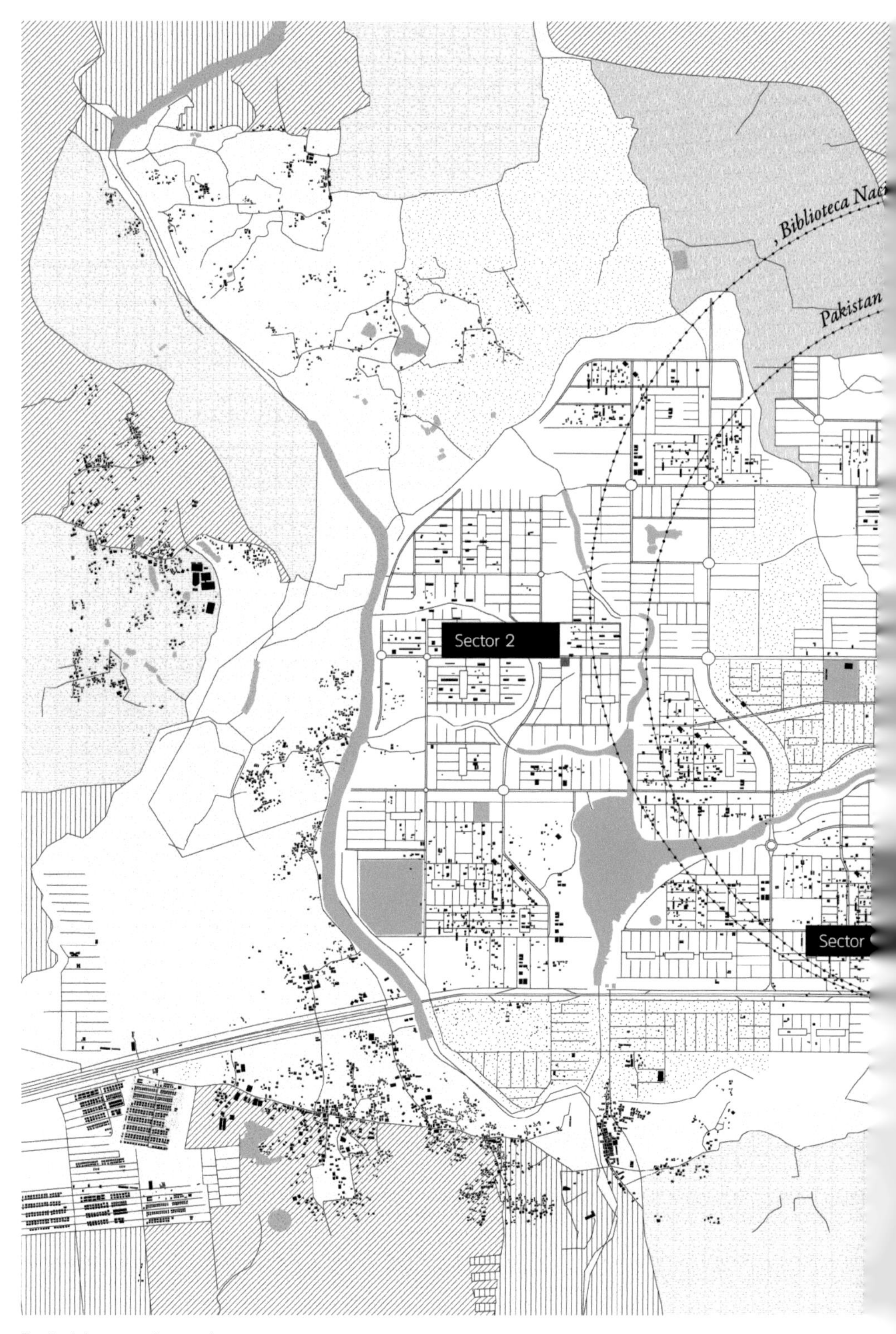

Territorial map on two scales.

ador, San Salvador, El Salvador
p Centre, Islamabad, Pakistan
Shitalskya River
50m
25
2km
1
0
Green Areas
Built Areas
Important Nodes
Infrastructural Nodes

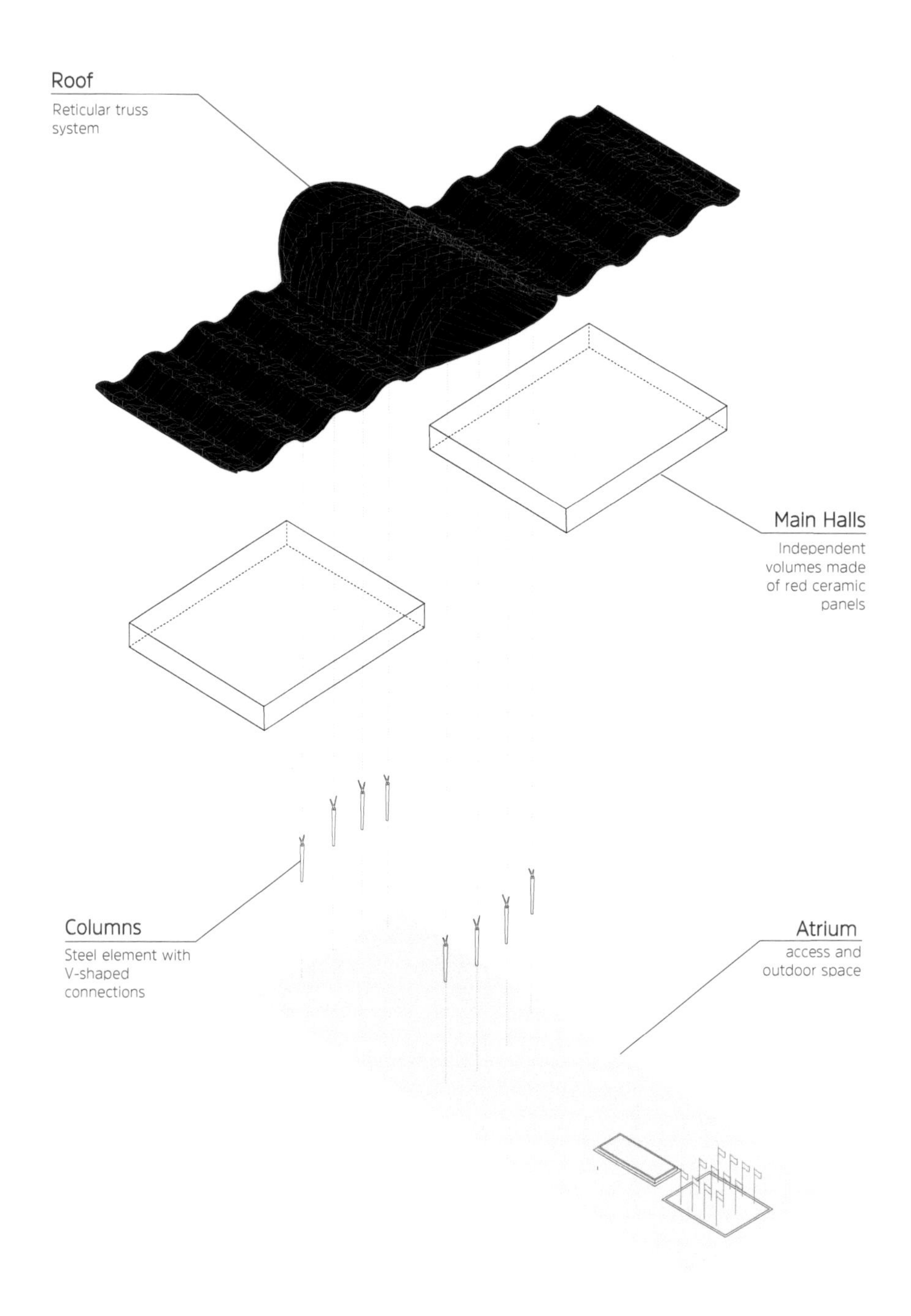

Relation diagram showing the main elements composing the tectonic of the building.

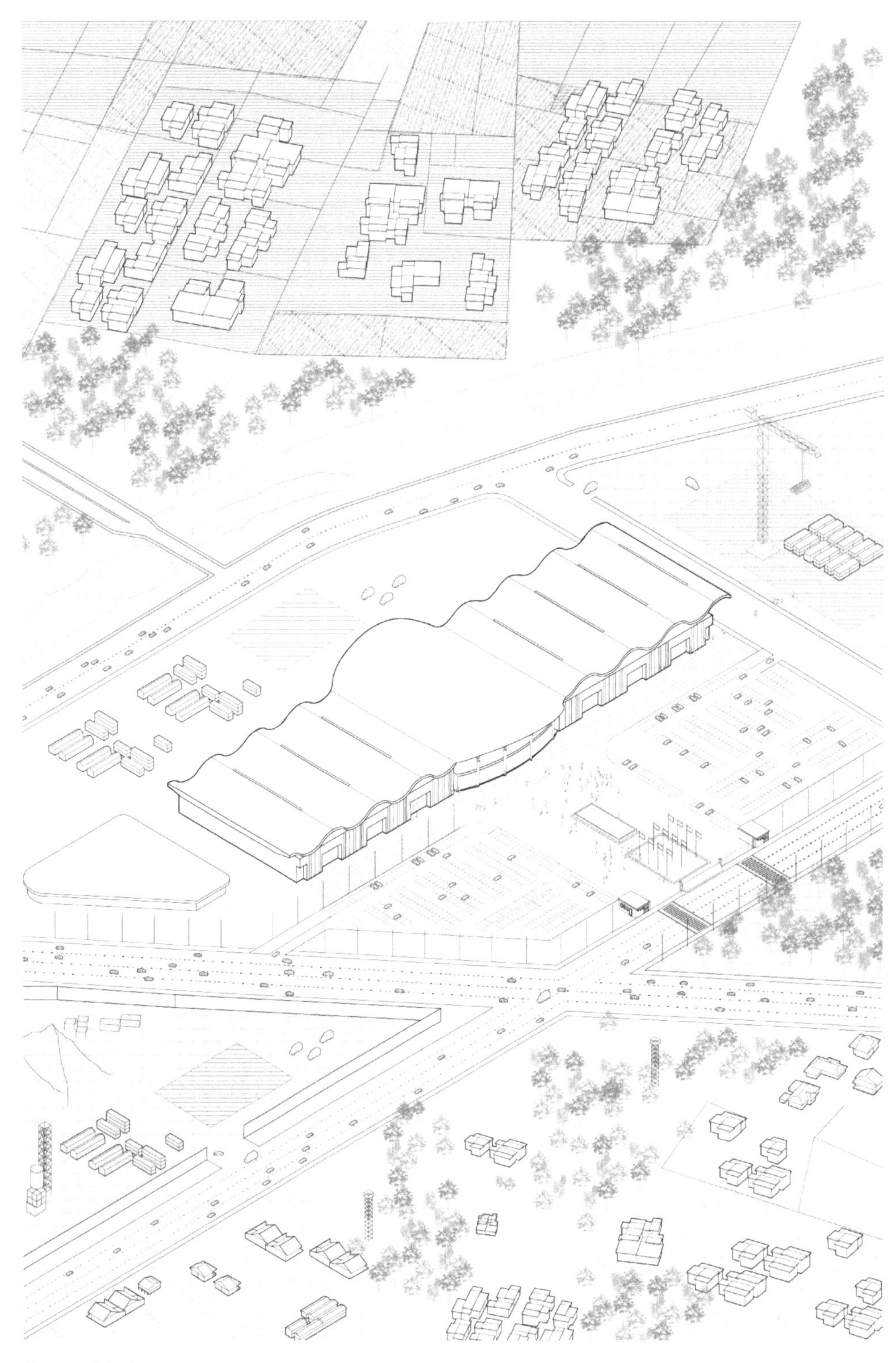

Axonometric view.

National Library of El Salvador

Location San Salvador, El Salvador
Floor Area 23,760 square meters
Date 2023
Architect Central-South Architectural Design Institute (CSADI)
Client National Government of El Salvador

The U-shape of the building and the horizontal louvers system on the façade resembles an open book with text lines on its pages.

The National Library of El Salvador, inaugurated on November 14, 2023, is located in San Salvador, the country's capital. This landmark project, located on the site of the former Francisco Gavidia National Library, symbolizes the first major collaborative venture between China and El Salvador pursuant to the establishment of diplomatic ties in 2018. In December 2019, Salvadoran president, Nayib Bukele announced the construction of a new national library following a state visit to China where he signed a memorandum of understanding with China's president, Xi Jinping within the framework of the Belt and Road Initiative. With an estimated budget of over 54 million US dollars, the project was to be implemented with Chinese cooperation.

The library has a total floor area of 23,760 square meters, of which seven stories are above ground and the others underground. Roughly 95 parking spaces, several logistics areas, and archive storage and technical equipment rooms are located below ground. Above ground, the building is made up of a raised squared podium, connected to the public area by a series of ramps and stairs, and an organic double-curved glazed volume above. It houses a comprehensive range of interconnected functional areas including: book collection zones, lending areas, public activity spaces, administrative offices, and a cafeteria.

The volume is U-shaped. The reading areas on each floor are arranged in the two wings around the big atrium in the middle of the U. The atrium is connected to a series of introverted outdoor spaces, situated between the two wings, and with a privileged view of the ancient cathedral of San Salvador located in front. The design also maximizes the influx of natural light entering through the tall mullion curtain walls protected by a system of shading devices in order to improve reading quality. The building's U-shape also has a symbolic purpose: it resembles an open book. The horizontal louvers system on the façade represents the text lines on its pages.

The National Library of El Salvador was the first Chinese foreign aid project to be completely digitally designed and delivered using digital twin technologies. The Central-South Architectural Design Institute, in cooperation with the general contractor, Yanjian Group Co. Ltd., developed the project using the Dassault 3DEXPERIENCE platform, a cutting-edge cloud-based application used to manage the design process from its earliest conception to construction on site. The software provided the architectural team with more freedom during the design development phase, especially as concerns the building's envelope. During the construction phase the software produced installation instructions for every single component and enabled on-site three-dimensional delivery with a laser pointer; this made it possible to build the double-curved façade without the need for specialized labor; local labor was used instead, thereby reducing costs and contributing to the local economy.

Territorial map on two scales.

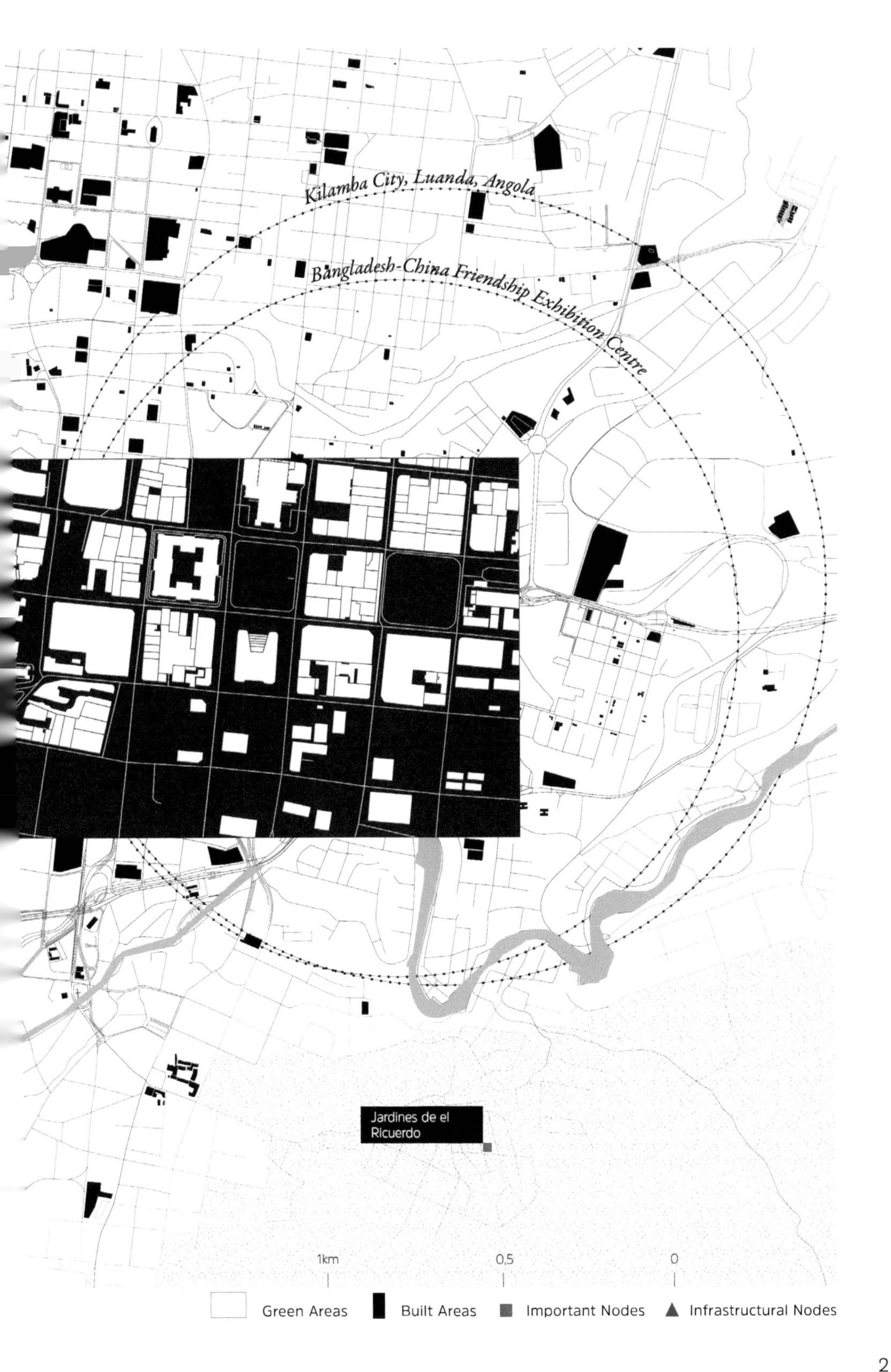
Kilamba City, Luanda, Angola
Bangladesh-China Friendship Exhibition Centre
Jardines de el Ricuerdo
1km
0,5
0
Green Areas
Built Areas
Important Nodes
Infrastructural Nodes

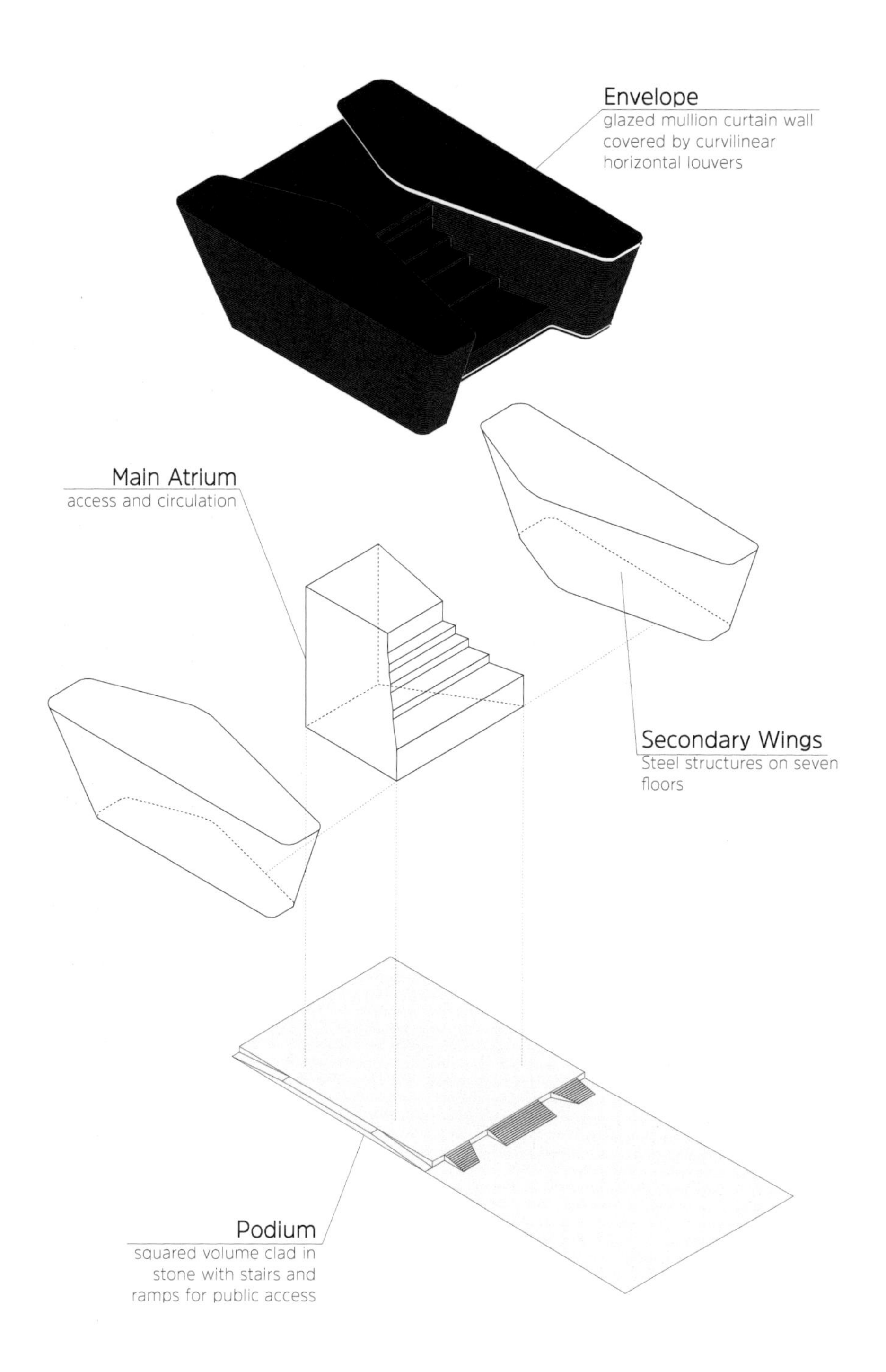

Relation diagram showing the main elements composing the tectonic of the building.

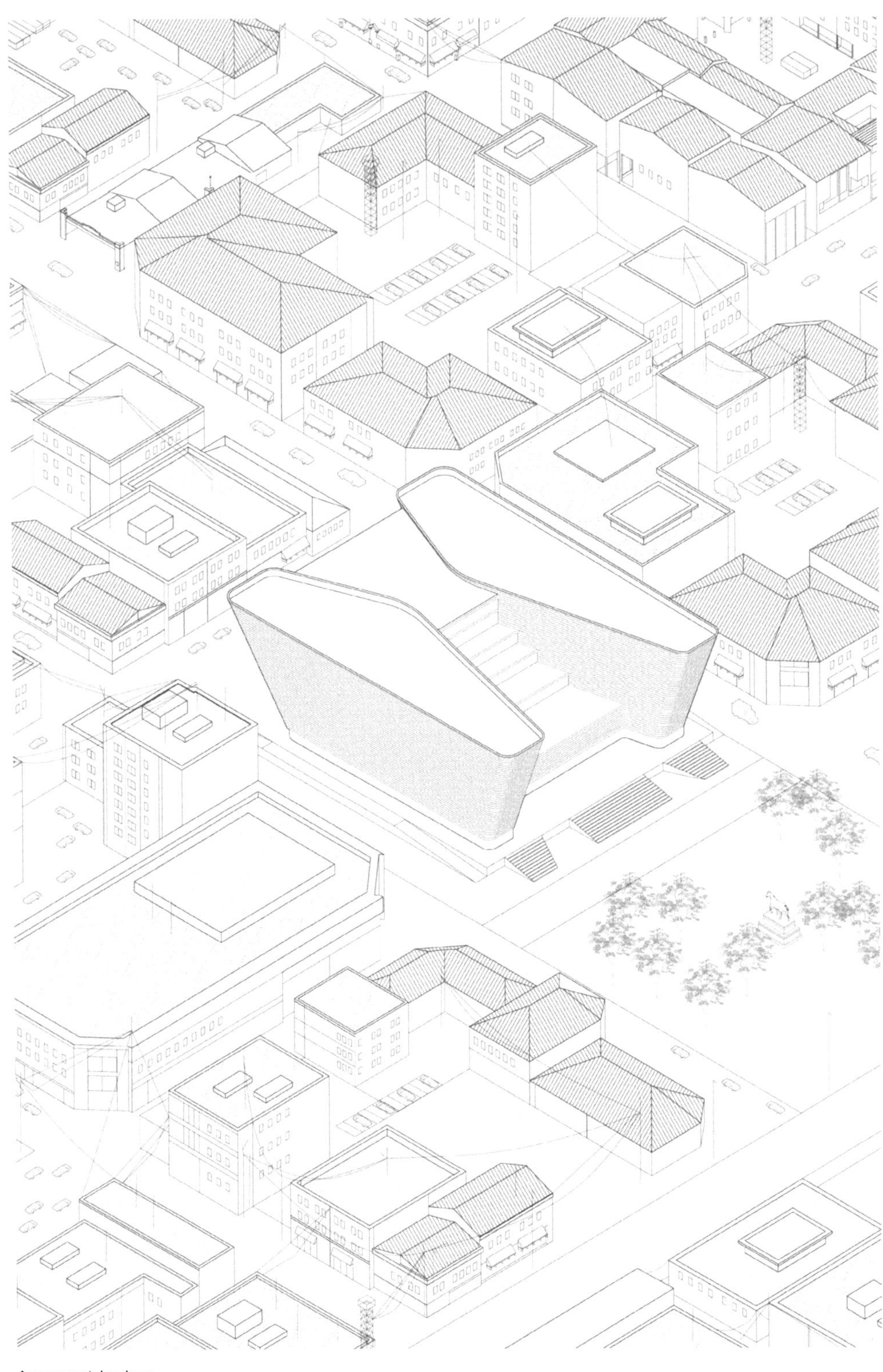

Axonometric view.

Bibliography

Ai, Xia. "18 Years of Overseas Design Practice of CCDI Group as a Complex Journey." In Charlie Xue and Guanghui Ding eds. *Exporting Chinese Architecture: History, Issues and "One Belt One Road."* Singapore: Springer, 2022.

Alexander, Christopher, and Howard Davis. *The Production of Houses*. Oxford: Oxford University Press, 1985.

Amoah, Louis. "China, Architecture and Ghana's Spaces: Concrete Signs of a Soft Chinese Imperium?" *Journal of Asian and African Studies 51*, no. 2 (2016): 238–255.

Apostolopoulou, Elia, Han Cheng, Jonathan Silver, and Alan Wiig. "Cities on the New Silk Road: The Global Urban Geographies of China's Belt and Road Initiative." *Urban Geography 45*, no. 6 (2024): 1095–1114.

Avermaete, Tom. "Coda: The Reflexivity of Cold War Architectural Modernism." *The Journal of Architecture 17*, no. 3 (2012): 475–477.

Avermaete, Tom. "Death of the Author, Center, and Meta-Theory: Emerging Planning Histories and Expanding Methods of the Early 21st Century." In Carola Hein, ed. *The Routledge Handbook of Planning History*. New York: Routledge, 2017.

Barnett, A. Doak. *Communist China: The Early Years, 1949–55*. London: Pall Mall Press, 1964.

Barthes, Roland. *Empire of Signs*. New York: Macmillan, 1982.

Bartke, Wolfgang. *The Economic Aid of the PR China to Developing and Socialist Countries*. New York: K.G. Saur, 1989.

Bélanger, Pierre. *Landscape as Infrastructure: A Base Primer*. Abingdon, Oxon; New York: Routledge, 2016.

Bonino, Michele, Francesco Carota, Francesca Governa, and Samuele Pellecchia, eds. *China Goes Urban. The City to Come*. Milan: Skira, 2020.

Bonino, Michele, Francesca Governa, Maria Paolo Repellino, and Angelo Sampieri, eds. *The City After Chinese New Towns: Spaces and Imaginaries from Contemporary Urban China*. Basel: Birkhäuser, 2019.

Bosker, Bianca. *Original Copies: Architectural Mimicry in Contemporary China*. Honolulu: University of Hawaii Press, 2013.

Botz-Bornstein, Thorsten. *Transcultural Architecture: The Limits and Opportunities of Critical Regionalism*. New York: Routledge, 2017.

Brautigam, Deborah. *The Dragon's Gift: The Real Story of China in Africa*. Oxford: Oxford University Press, 2011.

Bunnell, Tim. "BRI and Beyond: Comparative Possibilities of Extended Chinese Urbanisation." *Asia Pacific Viewpoint 62*, no. 3 (2021): 270–273.

Cai, Peter. *Understanding China's Belt and Road Initiative*. Sydney: Lowy Institute for International Policy, 2017.

Carota, Francesco and Giulia Montanaro. "Navigating the Nexus of Global DesignPractices and Local Construction Realties. An Exploration into the Collaborative Architectural Endeavors within the Framework of Belt and Road Initiative (BRI)." In Jiawen Han, Davide Lombardi and Alessandro Cece, eds. *Advances in the Integration of Technology and the Built Environment: Select Proceeding of AAB 2024*. Singapore: Springer Nature, in press.

Castells, Manuel. *The Informational City*. Oxford: John Wiley & Sons, 1991.

Charitonidou, Marianna. "Revisiting the Debate around Autonomy in Architecture. A Genealogy." In Federico Soriano, ed. *critic|all II International Conference on Architectural Design & Criticism. Actas Digitales, Digital Proceedings*. critic|all PRESS, 2016: 99–127.

Charleson, Andrew W. *Structure as Architecture*. Burlington: Elsevier, 2005.

Deleuze, Gilles. *Difference and Repetition*. London: The Athlone Press, 1994.

Delz, Sascha "Who Built This? China, China, China! Expanding the Chinese Economy through Mutual Benefit and Infrastructure Construction." In Marc Angélil and Dirk Hebel, eds. *Cities of Change: Addis Ababa. Transformation Strategies for Urban Territories in the 21st Century*. Basel: Birkhäuser, 2016.

Dikötter, Frank. *Mao's Great Famine – The History of China's Most Devastating Catastrophe, 1958–62*. London: Bloomsbury, 2010.

Dodson, Jay. "The Global Infrastructure Turn and Urban Practice." *Urban Policy and Research 35*, no. 1 (2017): 87–92.

Easterling, Keller. *Extrastatecraft: The Power of Infrastructure Space*. New York: Verso, 2014.

Easterling, Keller. *Organization Space: Landscapes, Highways, and Houses in America*. Cambridge, MA: The MIT Press, 1999.

Ernstson, Henrik, and Sverker Sorlin, eds. *Grounding Urban Natures: Histories and Futures of Urban Ecologies*. Cambridge, MA: The MIT Press, 2019.

Fei, Ding. "Worlding Developmentalism: China's Economic Zones Within and Beyond Its Border." *Journal of International Development 29*, no. 6 (2017): 825–850.

Frampton, Kenneth. *Studies in Tectonic Culture: The Poetics of Construction in Nineteenth- and Twentieth-Century Architecture.* Cambridge, MA: The MIT Press, 1996.

Franceschini, Ivan, and Nicholas Loubere. *Global China as Method.* Cambridge: Cambridge University Press, 2022.

Furlong, Kathryn. "Geographies of Infrastructure III: Infrastructure with Chinese Characteristics." *Progress in Human Geography 46*, no. 3 (2021): 915–925.

Gao, Hua, and He Lian. *Zhonghua Renmin Guoheguo Shi (A History of the People's Republic of China).* Beijing: Zhongguo Dangan Chuban She, 1995.

Gao, Yuan, Xue, Charlie, and Tan, Gang. "From South China to the Global South: Tropical Architecture in China under the Cold War." *Journal of Architecture 27*, no. 7–8 (2022): 979–1011.

Governa, Francesca, and Angelo Sampieri. "Urbanisation Processes and New Towns in Contemporary China: A Critical Understanding from a Decentred View." *Urban Studies 57*, no. 2 (2020): 366–382.

Governa, Francesca, and Angelo Sampieri. "Infrastrutture Globali e Divenire Urbano: Pireo, Trieste e il Corridoio Adriatico." *Territorio 103*, no. 4 (2023): 23–30.

Hays, Michael K. "Critical Architecture – Between Culture and Form," *Perspecta 21* (1984): 14–29.

Ibañez, Daniel, and Nikos Katsikis. *New Geographies 06: Grounding Metabolism.* Cambridge, MA: Harvard University Press, 2014.

Jabi, Wassim. "Digital Tectonics: The intersection of the physical and the virtual." *Proceedings of the 24th Annual Conference of the Association for Computer Aided Design in Architecture (ACADIA) and the 2004 Conference of the AIA Technology in Architectural Practice Knowledge Community* (2004): 256–269.

Jacoby, Sam, Alvaro Arancibia, and Lucia Alonso. "Space Standards and Housing Design: Typological Experimentation in England and Chile." *The Journal of Architecture 27*, no. 1 (2022): 94–126.

Kanai, J. Miguel and Seth Schindler. "Peri-Urban Promises of Connectivity: Linking Project-Led Polycentrism to the Infrastructure Scramble." *Environment and Planning A: Economy and Space 51*, no. 2 (2019): 302–322.

Kanai, J. Miguel., and Seth Schindler. "Infrastructure-Led Development and the Peri-Urban Question: Furthering Crossover Comparisons." *Urban Studies 59*, no. 8 (2022): 1597–1617.

Koolhaas, Rem. *Delirious New York: A Retroactive Manifesto for Manhattan.* New York: The Monacelli Press, 1997.

Koolhaas, Rem, Bruce Mau, Jennifer Sigler, and Hans Werlemann. *Small, Medium, Large, Extra-Large (S,M,L,XL).* New York: The Monacelli Press, 1995.

Lara, Fernando. *Global Apartments. Studies in Housing Homogeneity.* Raleigh: Lulu Press, 2009.

Latour, Bruno. *Pandora's Hope: Essays on the Reality of Science Studies.* Cambridge, MA: Harvard University Press, 1999.

LeCavalier, Jesse. "Human Exclusion Zones: Logistics and New Machine Landscapes." *Architectural Design 89*, no. 1 (2019): 48–55.

Lefaivre, Liane, and Alexander Tzonis. *Critical Regionalism: Architecture and Identity in a Globalized World.* Munich: Prestel Publishing, 2003.

Liu, Hu, and Huang, Ming. *Zhongguo duiwai yuanzhu yu guoji zeren de zhanlue yanjiu [The Strategy of China's Foreign Aid and International Responsibility].* Beijing: China Social Science Press, 2013.

Liu, Weidong, and Michael Dunford. "Inclusive Globalization: Unpacking China's Belt and Road Initiative." *Area Development and Policy 1*, no. 3 (2016): 323–340.

Lyster, Clare. *Learning from Logistics.* Basel; Berlin: Birkhäuser, 2016.

Marri, S. Ahmed. *Architecture for "Other." China's New Eclectic and Pragmatism in Developing Countries within the Framework of the Belt and Road Initiative.* Doctoral Dissertation, 2021.

Mao, Zedong. *Selected Works of Mao Zedong. Vol. 7.* Beijing: People's Press, 1996.

McNeill, Donald. "Skyscraper geography." *Progress in Human Geography 29*, no. 1 (2005): 41–55.

Mejía Hernández, Jorge, and Cathelijne Nuijsink. "Architecture as Exchange: Framing the Architecture Competition as Contact Zone." *Footprint 14/1*, no. 26 (2020): 1–6.

Mezzadra, Sandro, and Brett Neilson. *The Rest and the West: Capital and Power in a Multipolar World.* London: Verso, 2024.

Moe, Kiel. *Integrated Design in Contemporary Architecture.* New York: Princeton Architectural Press, 2008.

Murton, Greg. "Power of Blank Spaces: A Critical Cartography of China's Belt and Road Initiative." *Asia Pacific Viewpoint 62*, no. 3 (2021): 274–280.

Narins, Thomas P., and John Agnew. "Missing from the Map: Chinese Exceptionalism, Sovereignty Regimes, and the Belt Road Initiative." *Geopolitics 25*, no. 4 (2020): 809–837.

Newman, Winifred Elysse. "Tectonics in Equipoise." *TAD 7*, no. 1 (2023): 1.

Oakes, Tim. "The Belt and Road as Method: Geopolitics, Technopolitics and Power Through an Infrastructure Lens." *Asia Pacific Viewpoint 62*, no. 3 (2021): 281–285.

Oliveira, Gustavo de L. T., Greg Murton, Alberto Rippa, Thomas Harlan, and Yang Yao. "China's Belt and Road Initiative: Views from the Ground." *Political Geography 82* (2020): 1–4.

Olmo, Carlo. “One History, Many Stories.” *Casabella 59*, no. 619–620 (1995): 75–86.

Pevsner, Nikolaus. *Pevsner Architectural Guides. The Buildings of England*. London: Penguin Books, 1951.

Ponzini, Davide. *Transnational Architecture and Urbanism: Rethinking How Cities Plan, Transform, and Learn*. London: Routledge, 2020.

Pratt, Mary Louise. “Arts of the Contact Zone.” *Profession* (1991): 33–40.

Ramondetti, Leonardo. *The Enriched Field. Urbanising the Central Plains of China*. Basel: Birkhäuser, 2022.

Regilme, Salvador Santino F. and Obert Hodzi. “Comparing US and Chinese Foreign Aid in the Era of Rising Powers.” *The International Spectator 56*, no. 2 (2021): 114–131.

Reiser, Jesse and Nanako Umemoto. *Atlas of Novel Tectonics*. New York: Princeton Architectural Press, 2006.

Ren, Xuefei. *Building Globalization: Transnational Architecture Production in Urban China*. Chicago: The University of Chicago Press, 2011.

Rendell, Jane. “A Place Between Art, Architecture and Critical Theory.” Proceedings to Place and Location. Tallinn, 2003: 221–233.

Rendell, Jane. “Foreword”, in Jonathan Bean, Susannah Dickinson and Aletheia Ida, eds. *Critical Practices in Architecture: The Unexamined*. Newcastle upon Tyne: Cambridge Scholars Publishing, 2020: xi–xix.

Roy, Ananya. “What Is Urban about Critical Urban Theory?” *Urban Geography 37*, no. 6 (2016): 810–823.

Roy, Ananya, and Aihwa Ong, eds. *Worlding Cities: Asian Experiments and the Art of Being Global. Studies in Urban and Social Change*. Malden: Wiley-Blackwell, 2011.

Safina, Astrid, Leonardo Ramondetti, and Francesca Governa. “Rescaling the Belt and Road Initiative in Urban China: The Local Complexities of a Global Project.” *Area Development and Policy* (2023): 1–20.

Sanaan Bensi, Negar, and Francesco Marullo. “The Architecture of Logistics.” *Footprint 10* (2018): 1–8.

Schindler, Sebastian, Jason DiCarlo, and Deepak Paudel. “The New Cold War and the Rise of the 21st-Century Infrastructure State.” *Transactions of the Institute of British Geographers 47*, no. 2 (2022): 331–346.

Schultz, Anne-Catrin. “Architectural Tectonics in the Age of Climate Crisis, Social Change and Digital Fabrication.” TAD 7, no. 1 (2023): 2–3.

Shin, Hyun Bang, Yimin Zhao, and Sin Yee Koh. “The Urbanising Dynamics of Global China: Speculation, Articulation, and Translation in Global Capitalism.” *Urban Geography 43*, no. 10 (2022): 1457–1468.

Sidaway, James D., Simon C. Rowedder, Chin Yuan Woon, Weiquiang Lin, and Vatthana Pholsena. “Introduction: Research Agendas Raised by the Belt and Road Initiative.” *Environment and Planning C: Politics and Space 38*, no. 5 (2020): 795–802.

Silver, Jonathan “Corridor Urbanism.” In Michele Lancione and Colin McFarlane, eds. *Global Urbanism*. London; New York: Routledge, 2021.

Simone, AbdouMaliq, Dominique Somda, Giulia Torino, Mija Irawati, Nitin Bathla, Rodrigo Castriota, Simone Vegliò, and Tanya Chandra. “Inhabiting the Extensions.” *Dialogues in Human Geography 0*, no. 0 (2023).

Somol, Robert and Sarah Whiting. “Notes Around the Doppler Effect and Other Moods of Modernism.” *Perspecta 33* (2002): 72–77.

Speaks, Michael. “Design Intelligence: Part 1, Introduction.” *A+U* (2002): 10–18.

State Council. *White Book on China's Foreign Aid*. Beijing: State Council of China, 2014.

Summers, Tim. “China's ‘New Silk Roads’: Sub-National Regions and Networks of Global Political Economy.” *Third World Quarterly 37*, no. 9 (2016).

Summers, Tim. “Negotiating the Boundaries of China's Belt and Road Initiative.” *Environment and Planning C: Politics and Space 38*, no. 5 (2020): 809–813.

Tafuri, Manfredo. *Progetto e utopia: architettura e sviluppo capitalistico*. Rome: Laterza, 1973.

Till, Jeremy. *Architecture Depends*. Cambridge, MA: The MIT Press, 2009.

Tobey, Aaron. “Architecture at Sea: Shipping Containers, Capitalism and Imaginations of Space.” *Architecture and Culture 5*, no. 2 (2017): 191–212.

Toops, Stanley. “Reflections on China's Belt and Road Initiative.” *Area Development and Policy 1*, no. 3 (2016): 352–360.

Vale, Lawrence J. *Architecture, Power, and National Identity*. London; New York: Routledge, 2008.

Vegliò, Simone, Andrea Pollio, Francesca Governa, Jonathan Silver, and Elia Apostolopoulou. “A Dialogue on Global Infrastructure-Led Urbanization: Concepts and Reorientations.” *Dialogues in Human Geography* (forthcoming).

Venturi, Robert. *Complexity and Contradiction in Architecture*. New York: The Museum of Modern Art, 1977.

Venturi, Robert, Denise Scott Brown, and Steven Izenour. *Learning from Las Vegas*. Cambridge, MA: The MIT Press, 1972.

Watkins, Josh. "Spatial imaginaries research in geography: Synergies, tensions, and new directions." *Geography Compass 9* (2015): 508–522.

Wiig, Alan, and Jonathan Silver. "Turbulent Presents, Precarious Futures: Urbanization and the Deployment of Global Infrastructure." *Regional Studies 53*, no. 6 (2019): 912–923.

Williams, Joe, Caitlin Robinson, and Stefan Bouzarovski. "China's Belt and Road Initiative and the Emerging Geographies of Global Urbanisation." *The Geographical Journal 186*, no. 1 (2020): 128–140.

Winter, Tim. *Geocultural Power: China's Quest to Revive the Silk Roads for the Twenty-First Century*. Chicago: University of Chicago Press, 2019.

Xue, Charlie. *Building a Revolution: Chinese Architecture Since 1980*. Hong Kong: Hong Kong University Press, 2006.

Xue, Charlie. *World Architecture in China*. Hong Kong: Joint Publishing Ltd. Co., 2010.

Xue, Charlie. "Western Architectural Design in China (1978-2010)." *New Architecture 2* (2012): 18–25.

Xue, Charlie, and Guanghui Ding. *A History of Design Institutes in China: From Mao to Market*. London; New York: Routledge, 2019.

Xue, Charlie, and Guanghui Ding, eds. *Exporting Chinese Architecture: History, Issues and One Belt One Road*. Singapore: Springer, 2022.

Xue, Charlie, Guanghui Ding, Wen Chang, and Yao Wan. "Architecture of 'Stadium Diplomacy' – China-Aid Sport Buildings in Africa." *Habitat International 90*, (2019).

Xue, Charlie, and Xiao, Jun. "Japanese Modernity Deviated: Its Importation and Legacy in Southeast Asian Architecture since the 1970s." *Habitat International 44* (2014): 227–236.

Yeh, Emily T., and Elizabeth Wharton. "Going West and Going Out: Discourses, Migrants, and Models in Chinese Development." *Eurasian Geography and Economics 57*, no. 3 (2016): 286–315.

Young, Liam. "Neo-Machine: Architecture Without People." *Architectural Design 89*, no. 1 (2019): 6–13.

Yu, Hong. "Motivation Behind China's 'One Belt, One Road' Initiatives and Establishment of the Asian Infrastructure Investment Bank." *Journal of Contemporary China 26*, no. 105 (2017): 353–368.

Zhang, Yun. *Zhongguo Duiwai Yuanzhu Yanjiu 1950–2010 [Research of China Foreign Aid 1950–2010]*. Beijing: Jiuzhou Press, 2012.

Zheng, Helen Wei, Stefan Bouzarovski, Sarah Knuth, Mathaios Panteli, Seth Schindler, Kevin Ward, and Joe Williams. "Interrogating China's Global Urban Presence." *Geopolitics 28*, no. 1 (2021): 310–332.

Zhu, Jian. *Architecture of Modern China: A Historical Critique*. London; New York: Routledge, 2009.

Zukin, Sharon. "Reconstructing the Authenticity of Place." *Theory and Society 40*, no. 2 (2011): 161–165.

Illustration Credits

Cover photo by CreatAr Images

Pages 6–9, maps by Sofia Leoni

Pages 26–32, photos and drawings by Charlie Xue

Pages 68–73, photos by Al Yousuf

Pages 74–79, photos by Raul Ariano

Pages 80–87, photos by Ivo Tavares Studio

Pages 88–93, photos by CreatAr Images

Pages 97–223, illustrations, maps, diagrams, and drawings by Sofia Leoni

Page 98, photos by Davide Monteleone

Page 104, photo by Arch-Exist Photography

Page 110, photo by Liu Chen

Page 116, photo by Ai Yousuf

Page 122, photo courtesy of SCG

Page 130, photo by Raul Ariano

Page 136, photo retrieved from https://www.prologis.cn/en/industrial-properties/spec/chongqingxibuguojiwuliuzhongxin

Page 142, photo retrieved https://en.orda.kz/new-old-faces-who-is-khorgos-visitors-now/

Page 148, photo courtesy of Great Stone Industrial Park

Page 154, photo retrieved from https://english.news.cn/20230821/62ce0b745d0e4c8ea944f552803ba10c/c.html

Page 162, photo retrieved from http://www.lzxq.gov.cn/system/2019/08/20/030005540.shtml

Page 168, photo retrieved from https://global.chinadaily.com.cn/a/202112/17/WS61bbf003a310cdd39bc7bf2a.html

Page 174, photo retrieved from https://geoln.com/georgia/tbilisi/12451

Page 180, photo courtesy of Raiwai Flats estate. Public Rental Boards, Fiji

Page 188, photo by Paulo Moreira

Page 194, photo by CreatAr Images

Page 200, photo by Christian Gahl/gmp Architects

Page 206, photo by Lidia Preti

Page 212, photo courtesy of Financial Express.

Page 218, photo retrieved from https://mp.weixin.qq.com/s/C3FWAHL_4F95k009ARkyFw

Every reasonable attempt to secure permissions for the visual material reproduced herein has been made by the authors. The authors apologize to anyone who has not been reached.

Outlook

This is not a conclusion. The research portrayed in the previous pages was conceived as a preparatory investigation through the routes of the Belt and Road Initiative, to show the effects of this phenomenon in architectural terms. Our work is meant to be an invitation to increase the number of studies in the field of transnational practices from the Global East to the Global South. Chinese architects, real estate developers and construction firms have completed a huge number of projects in the world, and this book has explored only a few of them, so it would be desirable to examine the phenomenon on a much broader scale.

This book is intended to open a living platform, a collective work that aims to map the emerging architecture of the Belt and Road Initiative and further contribute, in this way, to display its effects on the global built environment. We launch a call to scholars and practitioners from all around the world, to provide new case studies and interpretations. Even though, starting from our preliminary selection of case studies, categorizations and theoretical reflections have been made in this book, we look for further evidence of BRI architecture that follows these points:
Neutral. The BRI architecture cannot be resolved in dialectic and binary terms. It challenges dichotomous understandings of architecture that include the relationship between meaning and form, between material and digital, between local adaptation and standardization, and between autonomy of the discipline and its dependency on given contingencies.
Collective. The BRI architecture questions the figure of the author in the production of architectural outputs, prioritizing collective and cooperative mechanisms that involve a large number of stakeholders and communities.
Cosmopolitan. The BRI architecture is neither local nor global. It opens an architectural discourse grounded on different realities while keeping a transnational nature. For this reason, it necessitates the recognition of the autonomy and dignity of each participant that engages in a critical dialogue.

Indeed, the interpretations we documented in this book are not the sole variables that could be considered: given its intrinsic nature, characterized by practices of negotiation and diplomacy and a resulting sense of architectural neutrality, the production of BRI architecture helps to multiply the narratives.

Moreover, if this study, applying grounded analyses to few selected case studies, could have the merit to recognize emerging architectural issues, other methodologies and perspectives - including in-depth studies on construction and management organizations, structured data analyses, ethnographic researches - could probably be adopted to systematize the concepts we proposed. Further investigations could also lead to practice-relevant findings to approach transnational design projects in the future.

For the ones interested in contributing to this living platform for BRI architecture, more information can be found by scanning the QR code.
https://www.newsilkroad-arch.com/

Acknowledgments

This book is one of the outcomes of a broader research project funded by the Italian Ministry of University and Research (MUR), titled "Rescaling the Belt and Road Initiative: urbanization processes, innovation patterns and global investments in urban China." The research involves scholars from the Politecnico di Torino (architects, planners, economic and political geographers), and the Università di Macerata (urban economists and experts in Chinese philosophy and culture). The authors wish to express their profound gratitude to the coordinators of this research project, Francesca Governa (Politecnico di Torino - DIST) and Francesca Spigarelli (Università di Macerata), as well as to all the other scholars involved: Francesca Frassoldati, Leonardo Ramondetti, Astrid Safina, and Angelo Sampieri.

The book is the result of a long series of conversations among the authors (Michele Bonino, Francesco Carota and Sohrab Ahmed Marri) and a series of thesis seminars involving both master's and Ph.D. students, coordinated and organized by Michele Bonino and Francesco Carota between March 2021 and July 2022 at the Politecnico di Torino. The authors would like to acknowledge Jiayin Feng, Daria Fossa, Lara Giordano, Sofia Leoni, Stefano Mondozzi, Lidia Preti, Andong Xu, Liushi Xue, Xiaoxiao Zhu for their contributions in sourcing materials and preparing the graphics.

Although the contents of the book can be attributed to this collaborative effort, the introduction was written by Francesco Carota and Michele Bonino; *I. The Architecture of the Belt and Road Initiative: A New Architectural Order?* was written by Michele Bonino, Francesco Carota, and Sohrab Ahmed Marri; *II. Architectural Guide to the Belt and Road Initiative* was curated and edited by Francesco Carota, with all the short essays in the "Gift Complexes" and "Mass Housing Enclaves" sections written by Lidia Preti, while those in the "Worlds of Special Rules" and "Super Gathering Places" sections written by Stefano Mondozzi, except for the essay on the National Library of El Salvador, which was written by Giulia Montanaro.

Finally, this book would not have been possible without the valuable feedback, reviews, and suggestions from the Birkhäuser copy editor and project manager, Ria Stein. The authors sincerely thank her for her dedication and contributions, which greatly improved this publication.

About the Authors

Michele Bonino, Ph.D. in History of Architecture, is Professor of Architecture and Urban Design and Head of the Department of Architecture and Design at the Politecnico di Torino. He was formerly Vice-Rector for international relations with China and Asian countries. His field of research is the innovation of design and its practices according to models of transnational exchange. Among other publications, his book *The City after Chinese New Towns. Spaces and Imaginaries from Contemporary Urban China* (with Francesca Governa, Maria Paola Repellino, and Angelo Sampieri) was published by Birkhäuser in 2019, and was also based on research at the China Room.

Francesco Carota, Ph.D. in Architecture, History, and Design from Politecnico di Torino, is Assistant Professor of Architecture at the University of Kansas. He is an associate member of the Center for East Asian Studies at the University of Kansas and an affiliate member of the China Room Research Group at the Politecnico di Torino. He is Co-founder and Principal of the multidisciplinary design firm Calibro Zero.

Francesca Governa, Ph.D., is Professor of Economic and Political Geography at the Interuniversity Department of Regional and Urban Studies and Planning (DIST), Politecnico di Torino. She is Director of the China Room Research Group at the Politecnico di Torino and a member of the Beyond Inhabitation Lab.

Sofia Leoni is an architect and Ph.D. candidate in Urban and Regional Development at Politecnico di Torino and has been part of the China Room Research Group since 2022. Her current research explores the impact of e-commerce platforms on rural China, highlighting the role of infrastructure in reshaping the contemporary countryside.

Sohrab Ahmed Marri, Ph.D. in Architecture, History, and Design from Politecnico di Torino, is Assistant Professor of Architecture at the Balochistan University of Information Technology, Engineering, and Management Sciences, Quetta. His research focuses on cross-cultural architecture, transnational design models, architectural exchanges under development aid, and urban issues in developing countries.

Stefano Mondozzi is an architect currently based in the Netherlands. He started a collaboration with the China Room Research Group in 2021. His master's thesis, carried out between the Politecnico di Torino and the École Polytechnique Fédérale de Lausanne (EPFL) in Lausanne, explores the impact of urbanization and logistics on architectural formation through the analysis of China's Belt and Road Initiative.

Giulia Montanaro is a building engineer and Ph.D. candidate in Architecture, History, and Design at the Politecnico di Torino. Currently, she is enrolled in a joint program between the Politecnico di Torino and Tsinghua University. She is an active member of the China Room Research Group at the Politecnico di Torino.

Lidia Preti is a Ph.D. candidate in Architecture, History, and Design and a member of the joint Ph.D. research program "Transnational Architectural Models in a Globalized World" between Politecnico di Torino and Tsinghua University in Beijing. Her research focuses on rural revitalization practices in vontemporary China, with a specific interest in the engagement of practitioners and academic institutions in these processes.

Charlie Xue is Professor of Architecture at the City University of Hong Kong. His research interests are in Chinese architecture, transnational design, and high-density environments. An award-winning architect and writer, he has published seventeen books, forty book chapters, and more than 140 research papers in professional and international journals. His book *Hong Kong Architecture 1945-2015: From Colonial to Global* (Heidelberg: Springer, 2016) received an award from the International Committee of Architectural Critics (CICA) in 2017.

This publication was realized with the support of the Interuniversity Department of Regional and Urban Studies and Planning and the Department of Architecture and Design of the Politecnico di Torino.

Graphic design concept: Jenna Gesse

Layout, typesetting, cover: Francesco Carota, Heike Strempel-Bevacqua

Copy editing: Erika Young

Copy editing for the publisher and project management: Ria Stein

Production: Heike Strempel-Bevacqua

Paper: 135 g/m^2 Magno Volume

Printing: Beltz Graphische Betriebe GmbH

Cover Xi'an Silk Road International Exhibition Center, Xi'an, China; photograph: CreatAr Images

Library of Congress Control Number: 2024916676

Bibliographic information published by the German National Library
The German National Library lists this publication in the Deutsche Nationalbibliografie; detailed bibliographic data are available on the Internet at http://dnb.dnb.de.

ISBN 978-3-0356-2669-8
e-ISBN (PDF) 978-3-0356-2673-5 Open Access
DOI: https://doi.org/10.1515/9783035626735.

published by Birkhäuser Verlag GmbH, Basel
Im Westfeld 8, 4055 Basel, Switzerland
Part of Walter de Gruyter GmbH, Berlin/Boston

The book is published open access at www.degruyter.com.
Printed on acid-free paper produced from chlorine-free pulp. TCF ∞

Printed in Germany

9 8 7 6 5 4 3 2 1

www.birkhauser.com

Questions about General Product Safety Regulation
productsafety@degruyterbrill.com